A Jersey Cattle Guidebook

by Ohio Jersey Cattle Club

with an introduction by Jackson Chambers

This work contains material that was originally published in 1922.

This publication is within the Public Domain.

This edition is reprinted for educational purposes
and in accordance with all applicable Federal Laws.

Introduction Copyright 2017 by Jackson Chambers

Self Reliance Books

Get more historic titles on animal and stock breeding, gardening and old fashioned skills by visiting us at:

http://selfreliancebooks.blogspot.com/

Introduction

I am pleased to present another title in the "Cattle" series.

The work is in the Public Domain and is re-printed here in accordance with Federal Laws.

As with all reprinted books of this age that are intended to perfectly reproduce the original edition, considerable pains and effort had to be undertaken to correct fading and sometimes outright damage to existing proofs of this title. At times, this task is quite monumental, requiring an almost total "rebuilding" of some pages from digital proofs of multiple copies. Despite this, imperfections still sometimes exist in the final proof and may detract from the visual appearance of the text.

I hope you enjoy reading this book as much as I enjoyed making it available to readers again.

Jackson Chambers

THE OHIO JERSEY  NINETEEN TWENTY-TWO

JERSEY PUBLICITY

THE object in annually publishing "The Ohio Jersey" is to firmly establish favorable publicity for the Ohio Jersey breeder and his cattle.

We know of no more economical method than each year presenting the Jersey breeders at large with a permanent resume of up-to-date achievements in the Jersey World in order that they may keep fully posted and dispense the latest facts which keep the Jersey in the foremost ranks of the dairy industry.

It is hoped that the editorials contained herein will prove beneficial to our many readers, and that the recipients of this booklet will seriously consider those breeders and dealers whose advertisements have helped to make it possible for us to issue it. We know the people placing their advertisements in this book can be relied upon at all times for a square deal, and we will greatly appreciate it, if when writing them, you will mention this booklet.

The Ohio Jersey Cattle Club does not undertake to sell Jerseys but functions largely as a medium through which the breeders voice their sentiments and co-operate for the best interest of the breed.

All enquiries for Jerseys are referred to the local clubs or individual breeders having the class of cattle called for. Our secretary keeps well informed as to what communities have surplus stock and their blood lines.

It has been said, and justly so, that Ohio Jersey breeders and their cattle have attained a high sphere in the Jersey World, this being largely brought about by harmonious dealings with their fellow men, and has resulted in the knowledge that wherever, and whenever Jersey cattle are mentioned, it is always conceded that "Ohio is the best place to buy good Jerseys."

I HAVE AT ALL TIMES A FEW CHOICE

JERSEYS

TO SELL AT REASONABLE PRICES.

JOE MORRIS
WESTERVILLE, OHIO

Keep This Type in Mind

COME TO OHIO FOR GOOD JERSEYS —— THREE THOUSAND BREEDERS WAITING TO SERVE YOU.

OHIO

L. J. TABER, Director of Agriculture.

WHEN the Divine Hand created the universe, a gem was carved out and later called Ohio. Between the lake and the river nature profusely spread a bounteous soil, a splendid climate, a location favorable to markets and distribution, which brought to Ohio a high type of pioneers.

Our soil is not only fertile, but wonderfully diversified. Ohio has as fine corn and grain lands as can be found. In Northwestern Ohio sugar beets are produced equal to the best irrigated lands of the west. Along the lake and rivers apples and fruits with flavor and quality rivalling the Hood River Valley are produced in abundance. In central Ohio and in the eastern section of the state general farming and stock raising are conducted in a character challenging national attention.

The wools of the State have a national reputation. The hill sections furnish as fine grazing land as the famous blue grass areas of Kentucky. Ohio is preeminent along many agricultural lines, but in dairying Ohio has given a definite challenge to the nation. More high producing cows can be found in Ohio than in any other state in the Union. The reasons are ample. The climate, the soil, the water, the markets, and lastly the citizenship of Ohio are ideally adapted for the Jersey cow.

The Buckeye State in addition to its agriculture has remarkable mineral resources, coal, iron, gas, oil, clay, lime stone, and other mineral products which enrich the commonwealth.

Between the lake and river beautiful valleys have been furrowed, down which winding streams are coursing with abundant water supplies for developing industry.

Across our soil pass most of the transcontinental transportation lines. Water plus transportation spells industrial development. Ohio is a state of cities, giving manufacturing and commerce a tremendous impetus. Diversified agriculture, diversified climate, diversified soil, transportation, and water power give Ohio an economic commercial life sound and progressive.

Ohio is not only a commonwealth noted for its agricultural, industrial, and commercial supremacy, but for the strength of its citizenship, and is the Mother of Presidents as well.

Come to Ohio to live, prosperity and happiness await you. Come to Ohio to buy your manufactured products and quality is guaranteed. Come to Ohio to buy live stock. Fou will buy health, vigor, production, reproduction, and prosperity.

TABLE ON SILOS
CAPACITY OF SILOS, IN TONS (Corn Silage)

Depth of Silo	Inside Diameter in Feet									
	10	12	14	15	16	18	20	22	24	26
20	26	38	51	59	67	85	105	127	151	177
21	28	40	55	63	72	91	112	135	161	189
22	30	43	59	67	77	97	120	145	172	202
23	32	46	62	72	82	103	128	154	184	216
24	34	49	67	76	86	110	135	164	195	229
25	36	52	71	81	91	116	143	173	206	242
26	38	55	75	85	97	123	152	184	219	257
27	40	58	79	90	130	102	160	194	231	271
28	42	61	83	95	109	137	169	205	243	285
29	44	64	87	100	114	144	178	216	256	300
30	47	67	91	105	119	151	187	226	269	315
31	49	70	96	110	125	158	196	237	282	330
32	51	74	100	115	131	166	205	248	295	346
34	56	80	109	126	143	181	224	271	...	...
36	61	87	118	136	155	196	243	293	...	...
40	70	101	138	160	180	228	282	340	...	...

The silo should be such a diameter that 2 inches of ensilage is fed daily.

Minimum amount of silage to be fed daily from silos of various diameters.

10 feet....520 lbs.	14 feet....1,015 lbs.	20 feet....2,075 lbs.
11 feet....625 lbs.	16 feet....1,325 lbs.	22 feet....2,510 lbs.
12 feet....745 lbs.	18 feet....1,680 lbs.	24 feet....2,985 lbs.

COME TO OHIO FOR GOOD JERSEYS —— THREE THOUSAND BREEDERS WAITING TO SERVE YOU.

THE OHIO JERSEY  NINETEEN TWENTY-TWO

One Reason Ohio Has the Most Active State Jersey Club in Existence

It is an undisputed fact that the Jersey breeders in Ohio have been exceedingly fortunate in selecting energetic, loyal and competent officers and directors for their State Association, and that through the members' hearty co-operation and their willingness to do things, has gained for Ohio its enviable position in the Jersey world.

Amongst other activities, when the dire need of funds existed, a call was made to all members of the Club to donate a good well bred heifer calf to be given to the State Club for sale, and the proceeds to be used in further promoting the Jersey interest in Ohio.

The result was that twenty-nine calves were donated and sold for $4,075. This, together with $460 donated at the time of the sale gave the Ohio Jersey Cattle Club a fund of $4,535 to take care of its obligations and established a good working capital for its future lines of promotional work.

At this time we take pleasure again in publicly thanking those who contributed to this sale and made it possible to further add to Ohio's established prestige.

Hugh W. Bonnell, Youngstown, Ohio.
Grant Chidsey, Brunswick, Ohio.
Geo. E. Kryder, McClure, Ohio.
M. E. and E. F. Pyle, Clarksville, Ohio
L. P. Bailey, Tacoma, Ohio.
Tom Dempsey, Westerville, Ohio.
L. J. Taber, Barnesville, Ohio.
Ada Long, Urbana, Ohio.
W. R. Spann & Sons, Morristown, N. J.
D. H. French, Newton Falls, Ohio.
J. S. Neer & Son, Mechanicsburg, Ohio
Tom White, Hooker, Ohio.
Haines & Cochran, Blanchester, Ohio.
J. H. Lewis & Sons, Carrollton, Ohio.
Finch Farm, Dayton, Ohio.
Jacob E. White, Greenfield, Ohio.
C. E. Harris & Son, Medina, Ohio.
C. W. Damon & Son, Brunswick, Ohio.
Hartman Stock Farm, Columbus, Ohio.
Est. J. W. Martin, New Concord, Ohio.
Rufus Cox, Manchester, Ohio.
H. E. Stokes, Waynesville, Ohio.
Allen Bailey & Son, Barnesville, Ohio.
Mahoning Co. Jersey Club, Canfield, Ohio
Columbiana Co. Jersey Club, Salem, O
J. F. Martt, Zanesville, Ohio.
Robert Wylie, Circleville, Ohio.
Clermont Co. Jersey Breeders' Assn., Milford, Ohio.
Col. D. L. Perry, Columbus, Ohio.
C. C. Folck, Yellow Springs, Ohio.
Joe Morris, Westerville, Ohio.
R. H. Brown, Indianapolis, Ind.
Hancock County Jersey Club.
Springfield Jersey Cattle Club.

In addition to the calves given by the above, donations were made by the following:

Frank J. Kahler, Plain City, Ohio.... $50
Robinson Bros. & Clark, Plain City, Ohio............................ 50
Belmont County Jersey Club 50
Sandy Valley Jersey Club 50
Randall Anderson, West Austintown, O. 25
Delaware County Jersey Club 25
J. L. Sicker & Son, Coshocton, O....... 25
Westerville Jersey Cattle Club, Westerville, Ohio 25
Licking County Jersey Club 25
Erie County Jersey Club 25
Springfield Jersey Club 25
Central Ohio Jersey Club 25
Frank Casey, Cambridge, Ohio 25
Miss H. Anna Quinby, Columbus, O.... 25
W. E. Reinhart, Magnolia, Ohio 10

Other State Clubs can, with much profit to themselves, follow in Ohio's progressive foot steps if they would do some permanent good for the breed as a whole.

Memorize it—
"Ohio is the Best Place to Buy Good Jerseys."

COME TO OHIO FOR GOOD JERSEYS — THREE THOUSAND BREEDERS WAITING TO SERVE YOU.

SERVICE

The Aim of The Ohio Jersey Cattle Club

By HUGH W. BONNELL, President,
Youngstown, Ohio.

The officers and directors of the Ohio Jersey Cattle Club are striving to perfect an organization that will be able to do some vital service for every Ohio breeder.

The word Service has such a far reaching significance that there is hardly a need that can be thought of which will not come under this head.

For the present, at least, it seems that the Ohio Club cannot and should not be a selling agent. This function falls logically and naturally to the County Jersey Clubs, which are rapidly springing to life and activity in all the counties of Ohio.

The Jersey breeder, particularly the isolated one, is lost without the County Club. Here is where they must gather to transact all their local business and solve their problems of testing, both for tuberculosis and butter fat; to make plans for the public showing of their animals; to consult with each other and arrange for the sale of their products and their surplus stock. The chance for local work and improvement is unlimited and the social events of no less importance.

These local clubs and their meetings are a source of mental and moral uplift to the Jersey farmer that cannot be ignored and that we can no longer do without.

In just the same measure, the Ohio Jersey Cattle Club cannot do without the County Clubs and we look forward to the day, in the near future, when every County Club will join the Ohio Club in a body. When they do, our financial difficulties are over, we will all be working and pushing in the same direction and the results could hardly be overestimated.

Going one step farther we may safely say that without the State Jersey Clubs, the American Jersey Cattle Club could accomplish little. As it is, we have the strongest National Breed Association on earth and its power for good to the Jersey cow and Jersey breeders would know no bounds were it backed up by a strong State Club in every State in the Union.

Supposing we begin with the American Jersey Cattle Club at 324 W. 23rd St., New York City. How futile would be its existence with no State organizations behind it. The State Clubs are equally powerless without strong functioning County Clubs and the County Clubs would not exist except for the individual breeders.

It all begins and ends with the individual and depends entirely on each of us to do our part in standing back of and pushing the wonderful breed we have chosen.

Ten Leading States for Registrations of Jersey Cattle As Recorded by the American Jersey Cattle Club During 1920-'21

State	No. Breeders Registering 1920 1921	No. Animals Registered 1920-1921	No. Breeders Registering per 100 sq. m.	No. Animals Registered per sq. m.
VERMONT	45	1463	47.5	153.0
OHIO	1920	5041	46.6	125.7
TENNESSEE	990	3603	23.6	85.8
CONNECTICUT	112	380	22.2	75.3
MASSACHUSETTS	136	551	16.4	66.4
INDIANA	747	1874	20.5	51.5
NEW YORK	680	2241	13.5	44.5
PENNSYLVANIA	653	2039	14.5	45.2
NEW JERSEY	54	316	6.6	38.4
RHODE ISLAND	15	45	12.5	37.5

COME TO OHIO FOR GOOD JERSEYS —— THREE THOUSAND BREEDERS WAITING TO SERVE YOU.

FOX'S QUEEN OF MINERVA
14,403 pounds of milk, 823 pounds of fat.
First Ohio cow to make 800 pounds of fat.
Dam of one Gold Medal daughter and two others in the R. of M.

RALEIGH'S FENDORA
Gold Medal Cow—Twice a Silver Medal Cow
SENIOR FOUR-YEAR-OLD CHAMPION OF OHIO

———— : : ————

Young Bulls from the Blood
of These Cows for Sale

HUGH W. BONNELL
CRANBERRY RUN FARM

R. D. 4   YOUNGSTOWN, OHIO

Developing the High-producing Jersey Cow

By PROF. OSCAR ERF, Dairy Department,
Ohio State University, Columbus, O.

DEVELOPMENT NOT A SHORT TIME PROCESS It has been said that the principal factors in the completion of a high record are the cow, the man, and the feed. The cow inherits a large part of her ability to produce large amounts of milk, so before choosing a cow for forcing, she must have the breeding back of her that will warrant the time and effort to be spent on her.

It has been demonstrated time and again that the caretaker has a large influence on the cow's production. It is only necessary to cite the case of Carl Gockerell, who was able to make world's records, not only on one cow, but in different breeds. One authority estimates that a record is sixty per cent cow, and forty per cent man.

The greatest influencing factor that a man uses in making a record, is his intelligence in feeding. The terms "herdsman" and "feeder" have come to be almost synonymous, due to the fact that a record may be made or lost by the method of feeding. Also, the best herdsmen begin to fit an animal for test, not just before she freshens, nor as a yearling, but back to the days before she is born. It is in the womb of the dam that the first chance is given to a calf to develop into a high-producing cow.

CARE OF THE CALF BEFORE BIRTH Experiments have shown that the feed given to the dam before the birth of the calf has a direct influence on the health of the newborn calf. Some cows were fed only one grain, with it straw, the cows being divided into three groups: those fed oats and straw, those fed wheat and straw, and those fed corn and fodder. The resulting calves were poor, but increased in quality in these three respective groups. Another group was fed a well-rounded ration of mixed grains, and the calves were perfect. The ration should be balanced at all times, but it is even more important at the pregnant stage, because two lives, instead of one, are dependent on the feeding at this time.

Mineral feeds find an important place here. If a cow has access to good bluegrass pasture, the addition of potassium phosphate, potassium iodide, and iron sulphate is not so necessary as if no good pasture is available. Good leafy alfalfa hay also helps to take the place of minerals in mineral form. The signs of a lack of mineral feed in the dam are apparent in the calf. The resulting calf will be puny and undeveloped. Also, abortion, retained afterbirth, and a drop in the milk flow are evident in the cow herself. A seemingly good ration, and plenty of it, may be fed, but the deficiencies were not fed to the pregnant cow. A rash on the skin, a patchy hide, and general rundown condition are the results of a lack of mineral elements.

It must be understood that a definite feed or ration cannot be prescribed for every cow, because the cow must be treated as an individual. But to produce a healthy, well-grown calf, this ration might be used as a basis for feeding the pregnant cow. It should be fed during the dry period, and up until about two weeks before the calf is born.

200 lbs. Dried beet pulp
200 lbs. Wheat bran
100 lbs. Alfalfa meal
100 lbs. Cocoa bean meal.
                (Bulk ration.)
200 lbs. Wheat
200 lbs. Oats
200 lbs. Malted barley
200 lbs. Corn meal
 40 lbs. Cane molasses.
                (Carbohydrates.)
150 lbs. Linseed meal
300 lbs. Cottonseed meal
 50 lbs. Ground peanut kernels.
                (Oil and protein.)
 60 lbs. Mixture of salt, calcium carbonate, bone meal and blood meal.

Not more than eighteen pounds of silage should be fed at this time, and the grain mixture should be fed in just large enough quantities to keep the cow in good flesh. She should not be allowed to become fat.

The calf will be stronger if the dam is allowed to rest dry for a few weeks previous to calving. About two weeks before the calf is dropped, the grain ration may be cut down, and its place taken by a mixture of bran and dried beet pulp, which will act as a laxative and a cooling ration. The internal organs of the cow must be kept working normally if she is to drop a healthy calf without any after troubles. If the cow is not in a laxative condition, she should be given a dose of Glauber's salts each day for several days.

CARE OF THE NEW-BORN CALF If the calf is normal and strong, it will be up and sucking the cow in a few minutes after it is born. If it is not able to stand, it should be helped, because the first colostrum contains so many essential qualities. It should be left with the cow for twenty-four hours after birth, so that it is free to drink at will.

The salt in the colostrum acts as an antiseptic and purgative. The albumin congeals or curdles in the stomach, thus starting the paristaltic motion, and giving the stomach muscles some bulk to work on. Colostrum is a septicizer, because of the lactic acid, which kills off streptic infections.

Milk is the natural food for the calf, because it contains vitamines; being a liquid, it is easily taken; it is the simplest food in liquid form, to become a solid in the stomach; and it has regulating elements that start the digestive organs to functioning properly. Colostrum is much like blood. If used when fresh, it is highly beneficial, but if allowed to stand, it decomposes readily and forms toxins.

The best time to teach the calf to feed from a pail is at one day old. allowed to get used to sucking the cow is hard to break the calf of the habit, and teach it to drink.

For the first three weeks, the calf will drink about ten pounds of whole milk daily and should receive whole milk until at least this age. It is better to carry it on even further, until say, six weeks of age. Before this time, a little fine ground oats and bran may be sprinkled in the milk, or in the bottom of the pail. The milk should be gradually changed from whole to skim, and at the end of two months, the calf will be drinking from fifteen to eighteen pounds of skimmilk daily. The bran and oats will be helped by the addition of some bright corn and a little milletseed. The calf will not be eating more than a pound or a pound and a half daily, of the grain mixture. Potbellied calves are caused by overfeeding, and unsanitary feeding utensils. After the calf has learned to eat grain, it should be fed from a clean dry wooden box.

The calf can never be foundered if the partistaltic motion is unhindered, and the

---

THE WHITE STOCK FARM
TOM P. WHITE, Owner   **HOOKER, OHIO**

Breeder of Jersey Cattle, Big Type Poland China Swine, Percheron Horses

BEATALL'S LUCY PRINCESS 390783
Imported in dam by W. R. Spann & Sons.
R. of M., 306 lbs. butterfat as a senior yearling.

OXFORD'S MODEL GIRL 360881
Imported by W. R. Spann & Sons.
Register of Merit, 498 lbs. of butterfat as a 3-year-old.

My Herd Sire, SIGMA'S RALEIGH MAJESTY 156131, is sired by Violet's Majesty, a son of Royal Majesty; his dam is the Champion Dairy Cow of Iowa, Register of Merit record, 17,023.6 lbs. milk, 1011.6 lbs. of butter.

MAJESTY'S STAR, an extremely high class Majesty bull, is sire of a choice lot of heifers and bulls in my herd, out of splendid cows rich in the blood of Majesty, Golden Maid's Prince, Noble, Plymouth Lad, Financial King. For a High Class Son of Majesty's Star write me at once.

Also offer a choice lot of Big Type Poland China Pigs.

T. P. WHITE - - - - - **HOOKER, OHIO**

gastric juices are able to handle the feed. Most cases of calf-scours could be prevented by proper sanitation.

After three or four months, the calf should not be fed fine-ground grains, but the coarser product, so that the mascatory system can be developed. During the second six months of her life, the calf should receive from two to three pounds of grain daily, and an abundance of good mixed hay and not more than ten pounds of silage per day. This will be gradually increased until, at about eighteen months, she may be given the fitting ration, as fed to her dam before freshening.

DEVELOPING THE TWO-YEAR-OLD HEIFER The heifers can be fed bulky material to a larger amount than can the cow, because she is using her energy only to grow, while the cow must maintain herself, and at the same time produce milk. There is possibly a profit in feeding skim milk right back to young heifers, but it has not proved practicable, except where milk is cheap.

The mineral elements of the ration should be very closely watched with the growing heifers, because of the great need for minerals in growth. The heifer should be fed from six to eight pounds of the fitting ration, as prescribed for her dam, and her treatment will be such as to keep her quiet and unirritated.

CARE AT PARTURITION The freshening cow should have a light, clean box stall, in order to give her freedom to exercise all of her muscles. She should be allowed plenty of air, but should be shielded from drafts.

It must be remembered that sanitation at this time is a prime requisite. As soon as the cow has cleaned, the stall must be cleaned out and disinfected. The transfer of many diseases may be prevented by this precaution. At the first sign of milk fever, the udder should be inflated with air, and the cow raised so that her head is held up. This latter prevents her choking on her tongue.

Right after freshening, a little hay and some water may be given to her. Do not attempt to physic the cow. Physic should not be used until the bowels show a need for it. After danger of milk fever is past, part of the milk should be taken from her, if the calf does not take enough to relieve her. The swelling will have gone down after three or four days, and by this time the cow will be milked dry at every milking.

The feed should not be urged upon her, until she develops a natural appetite, but feed her beet pulp, ground oats, and bran, in a mash, for the first two or three days. At the end of this time, she may be gradually brought onto a ration that will give her enough for maintenance, and also to stand the strain of producing large quantities of milk. A basic ration, around which the feed could be built, is as follows:

600 lbs. Corn
500 lbs. Oats
400 lbs. Bran.
260 lbs. Oil meal
200 lbs. Cottonseed meal
40 lbs. Salt, bone meal, blood meal.

Distillers' grains or sweet gluten meal make a valuable addition to the ration. A pint daily of cane sugar molasses may be fed, and will furnish her with minerals, as well as carbohydrates. Barley or oats may be substituted for corn meal, which should never be fed alone. Cob meal is a waste of the cow's energy. The greater quantity of grain that is fed, the greater must be the concentration, but there must be enough fiber to act as a carrier of the waste material. This may be furnished in the beet pulp or bran.

One of the cardinal points in making a big yearly record is to sacrifice high early production, to keep up the production later in the lactation period. At all times, the feed should be accurately weighed, so that the feeder knows just what the cow is consuming. She should not be on full feed until three weeks after freshening, unless she is on seven-day test. The feed must be gradually increased in amount, from ¼ to ⅛ lb. daily increase. When the cow is nearing her capacity, which will be when she no longer shows a greedy appetite, the increase must be made more gradually. If the feed is increased too rapidly, the digestive fluids cannot take care of it, and a fermentation sets up in the intestine, and the cow goes "off her feed." If this happens, it is best to miss one feed entirely, and do not fail to give her a physic, or she will drop way down in production. She will not suffer this drop if the trouble in her intestines is remedied soon. Then, when she is started back on feed again, she should be started low, and again worked up to her full feeding capacity.

At this time, a stimulant may be used, such as nux vomica, gentian, aloes, digitalis, jaborandi, or arsenic. Stimulants should be given very sparingly, and very carefully, and immediately the cow has become normal again the feeding of stimulants should be stopped.

Water is one of the most necessary feeds, and will do much to keep the cow in the best condition. The amount of water a cow will drink also affects her milk production, and various systems are resorted to, to increase her water consumption. The feed should be moistened, but should never be sloppy.

(Continued on Page 18)

Ohio's Greatest Producing Herd

IS OWNED BY

THE HARTMAN STOCK FARM CO.

Columbus, Ohio

At present our Jersey Milking Herd consists of 125 Head.

During the month of May, 1922, we made

TWO WORLD'S RECORDS

Out of 40 cows on test in May, 35 of them were in the 50-lb. list.
35 Cows on Register of Merit test averaged 62.86 lbs. fat.
25 Cows on Register of Merit test averaged 67.18 lbs. fat.
(These above are world's records)
20 Cows on Register of Merit test averaged 69.83 lbs. fat.
15 Cows on Register of Merit test averaged 72.40 lbs. fat.
10 Cows on Register of Merit test averaged 75.50 lbs. fat.

These are our Herd Sires in Service:

SYBIL'S GAMBOGE OF WHITEHALL 170426, considered by prominent Jerseymen to be the best breeding son of Sybil's Gamboge. By Sybil's Gamboge 174663, out of Royal Belle, P. 18283, H. C., she by Pennithorpe's Raleigh 96185, he by Mabel's Raleigh, P. 3722.

ST. OUEN'S RAMSGATE 166424, by Imp. Ramsgate's Champion 93534, a son of Lucy's Champion; out of St. Ouen's Gamboge 247109, dam of Gamboge's Prince 105565, and a daughter of Gamboge's Knight.

GOLDEN MAID'S GOLDEN GAMBOGE 166694, by Gamboge's Prince 105565, a son of Imp. Golden Maid's Prince; out of Champion's Golden Maid 257725, a daughter of Lucy's Champion. Champion's Golden Maid is a granddaughter of Imp. Golden Maid's Prince.

BOMBAY'S RAMSGATE 182265, by Imp. Ramsgate's Champion 93534; out of Bombay's Maid 24716, a daughter of Golden Maid's Prince.

The Following Young Sires are for Sale:

GAMBOGE'S OISE 199548. Dropped Oct. 12, 1920. By Gamboge's Prince, out of Oise's Princess of H. S. F. 331313, test 447.28 lbs. fat, Class AA.

RAMSGATE'S PRINCE 198938. Dropped Jan. 18, 1921. By St. Ouen's Ramsgate 166424, out of Gamboge's Jessie of H. S. F. 379062, test 539.17 lbs. fat, Class AA.

GAMBOGE'S HANDSOME PRINCE 198939. Dropped Feb. 8, 1921. By Gamboge's Prince 105565, out of Czar's Handsome Lucy 274001.

SADIE'S RAMSGATE 198940. Dropped March 7, 1921. By Magenta's Ramsgate 166423, out of Gamboge's Sadie of H. S. F. 359386, test 352.24 lbs. fat, Class A.

POLLY'S TORMENTOR LAD 199443. Dropped March 14, 1921. By Gamboge's Prince 105565, out of Polly of H. S. F. 250620.

TORMENTOR'S NOBLE OXFORD 198941. Dropped April 14, 1921. By Goldie's Oise 166426, out of Prince's Emma of H. S. F. 350390, test 709.29 lbs. fat, Class AA, Gold and Silver Medal cow, junior 4-year-old champion of Ohio.

GAMBOGE'S JUBILEE PRINCE 198942. Dropped June 3, 1921. By Gamboge's Prince 105565, out of Exile Lady of H. S. F. 257787.

We also have a number of young bull calves for sale,
out of Register of Merit dams.

CERTIFIED MILK

By ADAM KRUMM, Manager,
The Hartman Stock Farm Co.,
Columbus, Ohio.

The production of Certified Milk involves numerous technicalities which do not present themselves in the production of the natural, raw or pasturized product.

To begin with, such milk is produced under the jurisdiction of the Milk Commission of the Academy of Medicine, who select a bacteriologist, a chemist, a veterinarian, and a medical inspector, who make examinations and inspections under their instructions at least once a month.

The paddocks and barns to which the cows have access, must be free from marsh or stagnant pools, crossed by no streams that may become contaminated and free from plants which may affect the quality of the milk deleteriously.

The barnyard must be free from manure and well drained, and the stable so located as to insure good drainage and at sufficient distance from other buildings, dusty roads or fields to avoid excessive dust, etc. The floors, gutters, mangers, and water troughs must be made of cement, properly sloped and constructed to provide perfect draining and allow for easy feeding and perfect cleaning. The stable must be large enough to insure six hundred cubic feet of air space for each cow and well lighted with an average of four or more square feet or window glass per cow, and so ventilated that no stale, disagreeable or strong odor is noticed on entering the building. There must also be means of artificial light as well as running water, wash basins, soap, brushes and clean towels for the milkers. The storing and preparing of the feed must be in a separate apartment and not above the cows. Manure must be removed at least twice a day, and gutters and stalls washed at least once a day. All cleaning must be done at least thirty minutes before milking, and the floors flooded and remain wet during the milking.

The milk building must be within easy access of the stable, but also at sufficient distance to allow of no dust or odors from the same. It must be divided into a wash room, provided with a high pressure steam sterilizer, hot and cold water, modern plumbing and efficient means of cleansing and sterilizing all utensils that come in contact with the milk, and a cooling and bottling room with an efficient, sanitary, enclosed cooler, and whatever appliances are necessary for the rapid and cleanly cooling of the milk.

The herd used in the production of certified milk must include no animal that is not absolutely free from tuberculosis or shows any evidence of acute, chronic, local, or other diseases. The entire herd must be tested twice a year. The cows must be cleaned, milked and fed regularly and always treated kindly, their udders and surrounding parts clipped.

At least an hour before milking begins, the cows are thoroughly cleaned with a moist brush and a moist cloth and compelled to remain standing until milked. Immediately before milking each cow, the udders and surrounding parts are thoroughly cleaned again with a clean, moist cloth, a separate cloth being used for each cow. We prefer hand milking machines as we find the use of machines has a tendency to a higher bacterior count.

The hands of the milkers must be thoroughly washed in water with soap and brush and well dried on a clean towel. Milking must be done at the same hours daily, each milker being dressed in a clean, wash suit. Five streams from each teat is drawn in a separate vessel and discarded. The remainder of the milk is then drawn into a steril pail.

As soon as the milk is received at the milk house, it is strained through a steril strainer composed of a double thickness of cheese cloth with a layer of absorbent cotton between, and immediately cooled to 45 degrees Fahr. This milk must not contain more than 10,000 per c.c., or less than 4 per cent fat and 12.5 solids. The milk must be absolutely free from impurities, coloring matter, thickeners or other foreign substances. It is then bottled in pint glass bottles and closed with a steril, paraffined cap bearing the date and the seal of the commission.

Certified milk is especially adapted for invalids or infant feedings as its content varies very, very little if any. It can be readily seen that there is more responsibility and greater obligations involved in the production and handling of certified milk on account of its very delicate character than in any other food product known and up to the present time it has never been produced profitably.

The founders of the Certified Milk movement have rendered this country a conspicuous service which has never been fully recognized.

SHEFFIELD FARM
(12 Miles from Cincinnati)
Glendale - - - Ohio
BREEDERS OF HIGH CLASS
REGISTER OF MERIT JERSEYS
Federal Accredited Herd

One of Sheffield Farm's Matrons
NOBLE'S OXFORD TULIP 372224

This great producing daughter of Enfield's Tulip 2nd, by Agatha's Oxford Noble, completed her year's test on July 27, 1922, producing

840.93 Lbs. Butter Fat (not yet authenticated)

We believe this cow on her next test will produce one thousand pounds of butterfat.
One of her sons is retained as herd sire.

**We Have Other Great Cows in Our Herd
Write Us When You Want Good Jerseys**

H. B. HARK - - - - - Manager

| THE OHIO JERSEY |  | NINETEEN TWENTY-TWO |

The Five Highest Cows in Each of the Eight Recognized Classes in the State of Ohio at the Time of Publication of This Booklet September 1, 1922

Under Two Years

Name and H. R. No. of Cows.	Owner at Time of Test.	Age.	Milk	Fat.
Beauty's Gold Medal 397975	L. J. Taber, Barnesville	1-11	10586	552.86
Lou's Carry of Fairview Farm 361964	The Fairview Farm Co., Geneva	1-10	8427	502.33
Marston Farm 555P. 229658	Byron E. Skillings, Springfield	1-10	7913	502.20
Lucy's Viola May 452702	J. I. Myers, New Dover	1-6	9779	479.44
Irene's Carrie 432533	J. W. Hall, Barnesville	1-11	9064	475.61

Two Years Junior

Sophie's Dolly Dimple 348582	Hugh W. Bonnell, Youngstown	2-3	10814	689.68
Lipsa 323967	R. L. Pike, Geneva	2-5	11509	684.32
Golden Fern's Lad's Treva 430630	Albert Leedom, St. Paris	2-3	11337	653.71
Lasifoso Pride 444859	M. C. Ebright, Shreve	2-4	10343	641.00
Sybil's Bay Girl 450649	Hugh W. Bonnell, Youngstown	2-1	9676	636.16

Two Years Senior

Chieftain's Flora 389619	Woodcliff Farm, Columbus	2-11	13179	689.03
Lad's Lady Riotress Irene 279715	Willis Whinery, Salem	2-8	12308	660.81
Sophie's Bertha's Sons Edith 513951	Willis Whinnery, Salem	2-6	13140	649.44
Ioa Queen 333655	The Fairview Farm Co., Geneva	2-9	11239	647.37
Fern's Bright-Eyed Girl 414364	W. I. Griffith, Galena	2-11	9969	622.04

Three Years Junior

Tycoon's Bayleaf 307042	Hal. H. Hill, Wickliffe	3-4	13990	681.79
Champion's Annie Laura 370398	Hugh W. Bonnell, Youngstown	3-4	12442	654.87
Raleigh's Rosetta 305982	I. R. Blackburn, Dayton	3-5	11845	639.95
Tycoon's Golden Fern 307045	Est. of Hal. H. Hill, Wickliffe	3-0	10117	638.01
Beauty's Gold Medal 397975	L. J. Taber, Barnesville	3-1	11939	623.72

Three Years Senior

Spot G. 2d 418325	A. C. Pidgeon, Beloit	3-6	12245	713.11
Oxford Vona Pride 396662	Earl Smith & Son, Westville	3-8	13664	687.92
Financial Raleigh's Foxy 345311	Hugh W. Bonnell, Youngstown	3-7	12116	667.25
Huckleberry Princess 351632	W. H. Sawyer, Columbus	3-6	10750	663.44
Ziffa 323966	The Fairview Farm Co., Geneva	3-9	11058	632.62

Four Years Junior

Prince's Emma of H. S. F. 359390	Est. of S. B. Hartman, Lockbourne	4-2	12318	709.29
Miss Melia Oxford 394807	L. P. Bailey, Tacoma	4-1	12392	707.92
Sultan's Pretty Lady 363640	Hugh W. Bonnell, Youngstown	4-5	11407	686.42
Gamboge's Golden Queen 363259	Homer B. Slagle, Poland	4-0	12711	679.35
Theo's Bright Beauty 277600	L. P. Bailey, Tacoma	4-2	12027	638.83

Four Years Senior

Lady Rushcreek's Lassie 391043	Charles H. Stemen, Johnstown	4-6	12392	815.39
Raleigh's Fendora 305980	I. R. Blackburn, Dayton	4-7	12554	679.79
Pleasant Hill Dairy Lady 333055	Homer B. Slagle, Poland	4-6	12388	658.38
Financial Raleigh's Lass 344421	Hugh W. Bonnell, Youngstown	4-7	13008	649.48
Eminent's Belle of T. 367451	H. W. Bonnell, Youngstown	4-11	11041	633.37

Five Years and Over

Lucky Farce 298177	A. W. Murphy, Cleveland	8-3	18014	938.75
Theo's Bright Beauty 277600	L. P. Bailey, Tacoma	7-0	16976	917.37
Oxford Majesty's Victorie 370609	E. S. Kelly, Yellow Springs	10-0	18135	906.70
Snip Wauger 2d 243040	Hugh W. Bonnell, Youngstown	10-9	14877	823.27
Fox's Queen of Minerva 245459	Hugh W. Bonnell, Youngstown	6-7	14403	823.16

Twelve Years and Over—(Class 8A)

Duke's Iris 235101	Sheffield Farm, Glendale	12-1	12958	660.83
King's Trilby 216519	Mrs. Amie C. Newell, West Mentor	13-8	11438	646.32
Queen Esther Fox 187139	Hugh W. Bonnell, Youngstown	14-0	10371	596.84
Lustrous Girl 235083	E. S. Kelly, Yellow Springs	13-2	8175	511.04
Beryl of Cedar Grove 3d 175475	Willis Whinery, Salem	12-6	10368	493.45

COME TO OHIO FOR GOOD JERSEYS — THREE THOUSAND BREEDERS WAITING TO SERVE YOU.

DIXIE FARM

W. C. SHEPHERD, Proprietor   R. T. SHEPHERD, Manager

— :: —

MAJESTY JERSEYS
Herd on the U. S. Accredited List

CHRISTENA'S BEAUTY GIRL 505924
The Kind We Breed

— :: —

Herd Sire:

BLACK FLOWER'S MAJESTY 147284, a son of Imp. Oxford Majesty 134090. Dam, Black Flower 247129, a Register of Merit daughter of Gamboge Knight 95698.

Junior Sire:

MAJESTY'S DIXIE KNIGHT 201295, a son of Black Flower's Majesty 147284. Dam, Gamboge's Christena 311388, a Register of Merit daughter of Gamboge Knight 95698.

Every now and then, some good young things of both sexes for sale.

— :: —

DIXIE FARM   -   -   -   Hamilton, Ohio

Making Butter on the Farm

By WILLIAM WHITE, U. S. Dept. Agri.

Buttermaking begins with the production of the milk. Good butter can be made only from good, clean-flavored cream. To obtain practically all the cream from the milk and have it in the best condition requires the use of a cream separator.

The thorough cleaning and sterilizing of all dairy utensils is essential to the production of butter of good flavor.

Cream for buttermaking should contain about 30 per cent butter fat. A gallon of such cream will yield about 3 lbs. of butter.

Cream should be kept as cold as possible until time for ripening, when it should be warmed to from 65 degrees to 75 degrees Fahrenheit and held at that temperature until a mild-acid flavor is developed.

A thermometer should always be used in order to know that proper temperatures have been obtained.

Cream that is overripe (too sour) makes poor butter.

The churning temperature should be such that (1) the churning will require from 30 to 40 minutes, and (2) the butter granules will be firm without being hard—usually from 52 degrees to 60 degrees Fahrenheit in summer and from 58 degrees to 66 degrees Fahrenheit in winter.

All churning utensils should be cleaned, scalded and cooled before they are used.

The churn should be stopped when the butter granules are the size of grains of wheat.

The butter, in the granular condition, should be washed twice with pure water at about the same temperature as the buttermilk.

Buttermilk must be washed out, not worked out.

Salt should be added at the rate of about three-quarters of an ounce to the pound of butter.

Butter should be carefully worked until the salt is evenly distributed and a solid, smooth body is formed. The best butter has a waxy body, a bright appearance, and, when a slab is broken, a grain like broken steel.

Overworked butter has a sticky, salvy body, a dull, greasy appearance, and a gummy grain. Its keeping properties are not so good as in properly worked butter.

Mottled butter is caused by the uneven distribution of salt.

Butter for market should be in prints, wrapped in parchment paper and inclosed in paraffined cartons.

SINCE the introduction of the creamery system of butter manufacture into the United States the practice of making butter on the farm has gradually decreased and a marked change has taken place in the marketing of the product. The farm-made butter of today, instead of being shipped to the large markets, is consumed very largely at home and in the near-by towns, or is shipped to renovating factories. In spite of the fact that on the large markets the creamery product has almost entirely supplanted dairy butter, more than half the butter in this country is made on the farms.

To produce good butter it is necessary to begin with a good, clean-flavored milk. In some sections of the country it is customary to ripen and churn the whole milk instead of the cream. That practice, however, is inadvisable, because it requires a high churning temperature, which injures the quality of the butter and causes a considerable loss of butter fat in the buttermilk. It is also liable to result in too much water in the butter. For those reasons only the churning of cream will be considered. It is just as essential to obtain cream under such conditions that it will be of equally good quality as the milk.

METHODS OF SEPARATING CREAM Cream may be separated from the milk by gravity or by a centrifugal separator. Gravity separation may be accomplished by the shallow-pan, the deep-setting, or the water-dilution method. The first two have been extensively used and are still in use where very few cows are milked. In the first method the milk is placed in shallow pans and set in a cool place for about thirty-six hours, usually in a cellar or a spring house, and sometimes in cold water, to permit the cream to rise. During that time the surface, as a rule, is exposed to the air and frequently the cream absorbs or develops objectionable flavors. The skimmilk resulting from the removal of the cream by this method usually contains 0.5 to 1.5 per cent butter fat; that is, one-eighth to one-third of all the butter fat in the whole milk. It is frequently sour also; its value for calf feeding is injured, and its use in the household limited.

By the deep-setting method the milk as soon as drawn from the cow is placed in a "shotgun" can, which is placed in cold water, preferably ice water, for twelve hours. Because of the quick cooling to a low temperature the cream rises more quickly and completely than in the shallow-pan method and is skimmed before its fresh, sweet flavor has been lost. The resulting skimmilk may contain as low as 0.2 per cent butter fat, though often nearer 0.5 per cent, and is sweet. If the milk is not placed in ice water immediately after it has been drawn the loss of butter fat is still greater.

The dilution of milk with water has been

(Continued on Page 41)

WHEN BETTER JERSEYS
Are Bred on the ISLAND

W. R. SPANN & SONS
WILL IMPORT THEM

———— :: ————

In All Our Years of Importing Cattle We Have Never Overlooked the Fundamental Principle Upon Which to Judge Our Breed, Namely:

"UTILITY AS WELL AS BEAUTY"

Hamptonne Grey Beauty

Imported and sold in our 1922 sale to Mrs. Henry James, Oakdale, L. I., New York, for $4200. This was the record price for the 1922 importers sale.
On the cover of this booklet you will see the likeness of Fauvic's Simone Belle, one of the great cows imported by us. We had the misfortune to lose her just prior to our sale this year.

———— :: ————

We have several choicely bred imported-in-dam bulls for sale at reasonable prices.

Every Transaction Must Be Satisfactory to Our Customers

W. R. Spann & Sons - Morristown, N. J.

Members of the American Jersey Cattle Club and the Ohio Jersey Cattle Club.

How to Prepare Jerseys for Show or Sale

PAUL SPANN of W. R. Spann & Sons, Importers, Morristown, N. J.

MOST every cattleman and feeder has a way of his own for fitting cattle for the fairs or a sale. But after all, I think that most of the Veterans use practically the same methods of fitting.

Preparation should be made at least a year ahead of time in order to have cattle look their very best. The cows should be bred so that they will calve as near sale time as possible. There is no doubt but that more bloom can be put on a cow at freshening time than at any other. Cows which are to be fitted should be fed very light on grain for a few months before the time real fitting begins in order to give their stomachs a rest and get them cooled out.

Always begin feeding lightly with the grains on which you intend conditioning your cattle about four months before the sale. It is a good plan to never change the kinds of grain from this time until the cattle are sold. Increase the amount of feed per cow until you have them looking their best. Then hold them in this condition. Be careful not to crowd them too much, or you will throw them off feed or get them past bloom before sale time. It is better to bring them along very slowly. The best grain feed to use is a mixture of wheat bran, rolled oats, gluten feed, hominy, meal, and linseed meal. Feed about a quart of well soaked beet pulp along with this grain to each cow at a feed. Timothy and clover mixed about half and half is the best hay for fitting. Be sure that you have good bright hay, free from mould or the cattle will not take on that glossy appearance. Keep clean bedding under cows at all times. All the rubbing in the world will not get the stains out of a cow's hide if the bedding isn't kept clean.

Blanket cattle three months before the sale with heavy blankets. This will shed out all of the old hair and sweat the dirt out of the hide. Blankets keep the dust from the hay, straw and grain from getting into the hair. They help in putting a gloss on the coat and in training the hair.

During the last three weeks only soft brushes and a man's hands should be used in grooming the bodies of the cows. Coarse brushes and currycombs scratch and coarsen the hides. Good hand rubbing cannot be bettered for softening the hide, shedding out old hair and putting on a shiney appearance to the hair.

Cattle should not be turned out in the sunshine while fitting as the sun burns and dries out the hair. Never turn cows on grass while fitting them, as they get a taste for grass and will not eat hay. Grass does not fatten as fast as hay. Animals should be led out for exercise. This makes them easy to show to customers and easily handled in the sale ring.

Use a wood rasp on the horns about a month before the sale, taking off all of the rough shell and shaping up the horns. The day before the sale go over the horns with glass and emery paper, making them perfectly smooth. Then a beautiful polish can be put on them by rubbing pumice stone mixed with sweet oil on the horn with the hands and shining with a cloth.

The appearance of the cattle can be made much neater by clipping ears, tails, udders and bellies the day before the sale. All switches should be washed and plaited on this day and combed out on the morning of the sale.

Milk out the cows on the afternoon before the sale. Don't over bag them. Many a good cow has been turned down in the show ring just because her owner bagged her until her udder was all out of shape. Besides there is danger of rupture if they are bagged too tight. If a cow is a good one she will show plenty of udder by being filled up from the afternoon before.

Water and beet pulp thoroughly soaked in water are the best and safest means of filling cows on the sale day. Water cows on the morning before the sale and get them to eat as much salt and salty food on this day as you can. Do not water in the afternoon. On sale morning feed lightly on grain, water and fill them up on soaked beet pulp. Watch the cows because some of them may eat or drink enough to over barrel. This will cause them to look uncomfortable, and they will not show to advantage.

Cows should now be led into the ring by a man who knows how to handle them if they are to bring what they are worth. It isn't always the cows or the customer's fault that a cow doesn't bring what she is worth.

Mineral matter in the feed is necessary for a cow's health and her regular breeding and calving.

Developing the High-producing Jersey Cow
(Continued from Page 9)

A cow is naturally a nervous creature, and it is surprising what results are secured with different methods of handling. Here again it is noticed that a large part of the production is due to the way in which she is handled. A cow can't be blamed for kicking, if her milker swears at her. No lady would stand for that,—and a cow is a lady.

The production will be materially increased if the cow is milked more than twice a day, especially if she is naturally a good producer. A cow that is giving thirty pounds daily on two milkings a day, should be milked three times. And a cow that will produce fifty pounds, being milked three times a day, should be given a chance at four milkings a day.

CONCLUSION A dairy cow will produce to the limit of her capacity, only if given every assistance from the herdsman. She will respond to kind treatment and judicious feeding, such as no other animal will do. Her development should begin before she is born, and should continue all during her early life,—a stunted calf will never be able to take advantage of care that comes too late. The constituents of the milk are taken from the blood, so the blood stream must be active, and well-supplied with food.

THINGS A DAIRYMAN SHOULD KNOW.
"Hoosier."

Every dairyman should know—
The cost of producing a gallon of milk.
The cost of producing a pound of butter fat.
The cost of feeding a cow one year.
The cost of labor in caring for one cow one year.
The number of pounds of milk each cow in the herd yields each year.
The number of dollars each cow's milk brings each year.
Which is the most profitable cow in the herd, and why.
Which is the poorest cow in the herd and why.
How many boarders there are in the herd.
How much feed each cow will consume during the feeding period.
Which is the best and cheapest feed.—Jersey Bulletin.

I Can Sell 1000 Good Registered Jersey Cows, Bred Heifers and Heifer Calves

During the Next Twelve Months
IF I CAN BUY THEM AT REASONABLE PRICES

Don't write me about any Bulls you have to sell, because

I Have Eleven High Class Bulls for Sale at Present

Don't waste your time, nor mine, writing about common stuff, because my customers require good cattle and I do not handle any other kind.

Write fully, stating how many you have to sell, their ages, breeding and price.

JOHN A. LEE
WESTERVILLE, OHIO
Yes, I can use a few good Grades—What have you?

Feeding and Developing the Dairy Cow

By PROF. OSCAR ERF, Ohio.

INTRODUCTION Feeding cows for maximum milk production differs from feeding as ordinarily practiced. Usually making a direct profit from milk production is the object which the feeder has in view and in this case the average feeder endeavors to keep the cost of production below the income from the product.

Feeding cows for high milk production is carried on by only a small percentage of the dairymen, only those who are specializing in breeding and who desire to make a profit from the sale of calves rather than from the sale of milk. The breeder who is feeding for high records must determine the cow's ability to consume large amounts of feeding and convert it into milk. He makes an effort to test out the capacity of the digestive system and the power of the secretory organs. When these animals have been discovered an endeavor is then made to propagate these characteristics to future generations.

The following rules apply to selecting, developing and feeding cows where maximum production is desired:

SIZE 1. Select the largest typical cows of the breed to be tested. This does not mean that small cows are not good producers, but other things being equal, the larger the cow, the greater are the opportunities for high records.

APPETITE 2. Select the cows that are most greedy and that are good feeders. Appetite should not be temporary or spasmodic, but constant.

SELECTING COWS 3. Test cows for one year with fair feeding and select only the high producers for future work.

4. Cows should be well fed during the first year and the grain ration gradually increased in connection with the roughage. The amount of grain consumed depends on the grain mixture. If it contains a quantity of bulk feeds such as beet pulp, bran or ground alfalfa, then the grain ration may represent as much as 80 per cent of the ration. If, however, the ration is highly concentrated, then the proportion of grain should be reduced to an amount comparable to the amount of nutrients that the ration contains. This is necessary in order to create an active digestive condition. However in no case should the concentrates form less than 50 per cent of the entire ration.

Formerly the preparation ration for the first year consisted of one-third oats, one-third old process oil meal and one-third bran, with alfalfa hay and silage mixed with molasses.

This is a good ration for a cow that has not been worked hard previous to this time and that has a good constitution. However, for the average cow, the ration is not sufficiently complete and does not supply the necessary constituents in large enough quantities. For this reason a new preparation ration is suggested and it is proving successful in cases where it is used.

The basic part of the ration has been used for nearly ten years, but the cocoa bean meal has been recently introduced. The ration can also be used for economical milk production.

	Percentage composition
Bulk Ration—	
Dried Beet Pulp.................	
Wheat Bran	
Alfalfa meal	30
Cocoa bean meal.................	
Starch and Sugar Ration—	
Wheat	
Oats	
Malted Barley	42
Corn meal	
Cane Molasses	
Oil and Protein Ration—	
Linseed meal	
Cottonseed meal	25
Ground peanut kernels...........	
Mineral Ration	
Soluble Blood flour..............	
Bone Calcium Phosphate.........	
Calcium Carbonate	3
Salt	

The high grade ration may be fed in dry form although better results may be obtained if the feed is moistened. It should not be wet or sloppy. From twelve to eighteen pounds of silage should be fed with the ration and from six to fourteen pounds of alfalfa hay. The amount depends upon the individual animal.

In cases where alfalfa hay is not available, clover may be fed providing it is not moldy or mow burnt. Sometimes it is preferable to feed timothy and clover hay mixed since this hay cures better and really contains more digestible nutrients than clover alone. However, not more than one-third of the crop should be timothy.

DRY COWS 5. While the above preliminary high-grade ration is used to develop cows previous to the test

(Continued on Page 79)

FARRELL FARM

HEADQUARTERS FOR
SYBILS ——— MAJESTYS
And Other
IMPORTED and ISLAND BRED JERSEYS

Every female in the herd is either on test or already has a creditable Register of Merit record.

Occasionally we reserve a young bull suitable for first class breeding purposes.

Write for Description and Prices
Accredited Herd No. 12073

——— :: ———

FARRELL FARM
HERBERT FARRELL, Owner

Sandusky - - - - - **Ohio**

RIDGEVIEW FARM

L. J. Powers & Son
Owners and Breeders of

High Class
Registered Jersey Cattle

Rich in the blood of Majesty, Lucy's Oxford Lad, Stoke Pogis, St. Lambert, Champion Flying Fox.

—o—o—

Young Stock for Sale
At All Times.

L. J. Powers & Son
HURON - - - **OHIO**

Federal Accredited Herd

OAKLAND STOCK FARM

Plain City - Ohio

—:—

Herd Sires:
IMPERIAL MAJESTY 105682
 Sire, Royal Majesty. His dam is a daughter of Morny Cannon.

ROWENA'S MAJESTIC NOBLE 177470
 Sire, Dairylike's Majesty, P. 5380, H. C.
 Dam, Rowena's Pet 451270, 600 lbs. fat.

For MAJESTYS
See Me

FRANK J. KAHLER

THE OHIO JERSEY  NINETEEN TWENTY-TWO

TUBERCULOSIS IN LIVE STOCK

(Extract from "Farmer's Bulletin 1069," Published by The United States Department of Agriculture.)

This article is presented to the Jersey Breeders at large so that they may be fully aware of the importance of co-operating with Federal and State authorities in the elimination of Tuberculosis in Live Stock.

HOW TUBERCULOSIS SPREADS FROM A DISEASED HERD TO A HEALTHY ONE

Tuberculosis may be introduced into a healthy herd by any of the following means:

By the addition of an animal that is affected with the disease; therefore animals should be purchased only from herds known to be free from tuberculosis, or from herds under supervision for the eradication of the disease.

By feeding calves with milk or other dairy products from tuberculous cows; this frequently occurs where the owner purchases mixed skim milk from the creamery, and feeds it to his calves without first making it safe by boiling or pasteurization.

By showing cattle at fairs and exhibitions; reports have indicated that numerous herds have become infected through mingling with infected cattle at shows or by occupying infected premises.

The shipment of animals in cars which have recently carried diseased cattle and which have not been disinfected properly.

Community pastures; pastures in which tuberculous cattle are allowed to graze are a source of danger.

In most cases the outward appearance of the animal bears no relation to the degree of infection. The disease frequently develops so slowly that in some cases it may be months or even longer before any symptoms are shown; therefore be on the safe side and have your herd tested.

PRESENT KNOWLEDGE OF TUBERCULOSIS

Probably no disease affecting either the human race or live stock is better known or has been the object of so much study as tuberculosis. Present knowledge of the disease is derived from many sources, including the work of eminent scientists who discovered its cause, and studies of the numerous ways in which it is spread, of the manner by which man and animals contract it, and the effects it produces.

The tuberculin test—the means of detecting tuberculosis—was devised in 1882 by the eminent scientist, Dr. Robert Koch. Thus the test has been known for more than a third of a century. The facts regarding it and other information presented in this bulletin are based upon long experience and results verified many times. The methods recommended to be used in the eradication of tuberculosis have been tried upon large numbers of herds and found to be effectual and practical.

EARLY ERADICATION IS MOST ECONOMICAL

Live-stock owners are earnestly requested not to wait until the States and Federal Government come into their localities to eradicate tuberculosis. It would not be possible indeed, at this stage to undertake to eradicate tuberculosis from the live stock of the United States solely through organized official forces established by the respective States and the Federal Government. The area over which tuberculosis has spread is too vast, the herds too numerous, and funds are insufficient for conducting the work on so extensive a plan even though trained veterinarians were available in sufficient numbers to do the work. Every live-stock owner should be a party to this campaign which has been inaugurated to eradicate tuberculosis. In almost every locality of the United States are veterinarians capable of rendering valuable services to live-stock owners in this great work, and the cost of eradicating is greatly reduced by combating the disease in its early stages. Yet even in badly affected herds, eradication can be undertaken with success. There are records of many herds, in which three-fourths of the animals were affected with tuberculosis, which eventually were freed from it and afterwards maintained in a healthy condition.

The extirpation of tuberculosis from live stock is important not only from an economic standpoint, but also because a considerable percentage of tuberculosis in the human family, especially among children, is positively due to the consumption of infected milk or other dairy products from tuberculous cows. It is eminently proper for the respective State governments to expend funds for the maintenance of tuberculosis sanitariums for the care of persons afflicted with that disease, and likewise it is extremely important to use vigorous measures to check the marketing of germ-laden milk. While it is true that proper pasteurization of milk destroys the living organ-

(Continued on Page 32)

COME TO OHIO FOR GOOD JERSEYS —— THREE THOUSAND BREEDERS WAITING TO SERVE YOU.

FOR THE BLOOD OF THE WORLD'S CHAMPION JERSEYS
Attend Our
SECOND ANNUAL SALE
TO BE HELD AT

Shady Nook Farm   -   Brunswick, Ohio

Thursday, October 19, 1922

The Sophie's Tormentor Family Has Produced

23½% of all Gold Medal sires.
33⅓% of all Silver Medal sires.
The world's champion Register of Merit sire.
The world's champion long distance cow, and have won
64 A. J. C. C. Gold Medals for large yearly production,
and many other cherished awards.

Our Offering is replete in the blood of the following noted sires:
POGIS 99th OF HOOD FARM 47th 162841. Sire, Pogis 99th of Hood Farm. Dam, Lass 68th of Hood Farm, R. of M., 754 lbs. butter.
LASS 89th OF HOOD FARM'S SON 165860. Sire, Pogis 99th of Hood Farm. Dam, Lass 89th of H. F., first prize National Dairy Show, R. of M., 814 lbs. butter.
TORONO OF BATH 115553. Sire, Hood Farm Torono. Dam, Figgis 62nd of H. F., R. of M., 471 lbs. butter.

WRITE FOR A CATALOG

E. C. ROBINSON & SON,	GRANT G. CHIDSEY,
Yellow Creek Jersey Farm,	Shady Nook Jersey Farm,
Copley, Ohio.	Brunswick, Ohio.

THE SOPHIE'S TORMENTOR FAMILY HAS PRODUCED:
TOM DEMPSEY, Sale Manager, Westerville, Ohio.
Both Herds Fully Accredited by the Federal Government

MEDINA JERSEY CATTLE CLUB

MEMBERS:

Arthur G. Abbott......Wadsworth, Ohio.	C. W. Damon & Son....Brunswick, Ohio.
Geo. F. Abbott....Chippewa Lake, Ohio.	H. P. Diefenbach......Brecksville, Ohio.
Carl Abbott..............Medina, Ohio	E. I. Ganyard & Sons......Everett, Ohio.
Beecher Baxter..........Medina, Ohio.	C. E. Harris & Son........Medina, Ohio.
Roy G. Bissell.............Seville, Ohio.	Bert Hodgeman..........Medina, Ohio.
H. E. Bourne..........Brecksville, Ohio.	H. C. Hulbert.............Seville, Ohio.
T. B. Brown, Jr...........Spencer, Ohio.	Clair M. Norton..........Seville, Ohio.
Siegel Brown.............Medina, Ohio.	E. C. Robinson & Son......Copley, Ohio.
Harold Burr.................Lodi, Ohio.	Grant E. Tillotson.....Brunswick, Ohio.
Grant G. Chidsey......Brunswick, Ohio.	Ervin L. Wilkinson....Brunswick, Ohio.

Our Breeders can supply you with both Production and Type

——— Come and See Their Stcok ———

Many cows in the R. of M. and others qualifying.

Do not miss our annual sales.

——— : : ———

G. F. ABBOTT, Pres.,	GLEN L. GANYARD, Secy.,
Chippewa Lake, O.	Medina, Ohio, R. 2.

Feeding and Managing the Dairy Calf

By W. K. BRAINERD and H. P. DAVIS.

FEEDING the cow well before calving insures a strong, healthy calf. The best time to wean the calf is after it takes the first milk. Early weaning makes it easier to teach the calf to drink.

Everything about the calf should be scrupulously clean.

Milk from infected cows or from a creamery should be pasteurized before it is fed.

Calves should be fed sweet milk of a uniform temperature and should always receive a little less than they desire.

All calves should be fed regularly; very young calves should be fed three times a day.

At first the calf is fed whole milk, the quantity being gradually increased. Skimmilk is substituted as soon as practicable, and if cheap is continued until the calf is six months old. Ordinarily the maximum quantity of skimmilk that can be fed economically is 20 pounds a day. When the calf is two weeks old, grain and bright, clean hay should be offered; the quantity fed should be increased as the calf's appetite demands.

Milk substitutes are not equal to milk, but give fair results when used with care.

Quarters must be clean and dry, with plenty of bedding.

Stanchions save milk and prevent the calves from sucking one another.

Water is necessary for calves.

Marks for identification should be plain without disfiguring the animal.

Calf diseases are largely the result of filth and carelessness. Prevention is cheapest and best.

Young dairy stock should have all the hay they will eat, and grain in proportion to weight.

The heifer should be bred to freshen when about two years old. Handling before freshening prevents shyness.

Fall calving usually gives best results.

The young bull should have an abundance of feed, plenty of exercise, and not be allowed too heavy service.

The foregoing points on feeding and management of the dairy calf are discussed somewhat fully in this bulletin.

POORLY nourished cows give birth to weak, puny calves, which are hard to raise. The feeding of the calf, therefore, begins before it is born. The food elements necessary for the development of the calf are taken into the stomach of the cow, digested, assimilated and transmitted to the calf through the umbilical cord, the connection between the mother and the calf. It is evident that if the cow does not receive food enough to keep herself in thrifty condition and at the same time develop her calf, both she and the calf must suffer. In endeavoring to raise good, thrifty calves many dairymen handicap themselves at the start by not properly feeding the pregnant cows. Such cows should have an abundance of palatable and succulent or juicy feed in order to insure good body flesh and healthy, thrifty condition at calving time. The calves will then be well developed, strong and sturdy and ready to respond normally to proper feed and care.

It is assumed that the calf is not to be raised by sucking the cow, but is to be fed by hand. The longer it sucks, therefore, the more difficult it will be to teach it to drink. On the other hand, the first (or colostrum) milk of the cow possesses properties which stimulate the calf's stomach and other digestive organs to action. Colostrum is nature's physic, and for this reason the young calf should always receive its mother's milk at first. The calf is sometimes weak at birth, and for this reason should have nourishment as soon as possible. It is usually easier to induce the calf to suck the cow than to try to make it drink from the pail. Because of these facts most dairymen prefer to let the calf remain with its mother for about 48 hours immediately after birth. An additional advantage of this practice is that the dam will carefully dry the calf by licking within the first few hours of its life. In the case of a weak calf or one that does not gain strength readily it may be best to allow it to remain longer than 48 hours, although under such circumstances it is sometimes difficult to teach the calf to drink, and serious trouble may result from its failure to obtain food.

TEACHING THE CALF TO DRINK It is desirable that the calf be in thrifty, vigorous condition when it is taught to drink. It should be kept without food for at least 12 hours, at the end of which time it will be hungry and will usually drink milk from the pail much more readily than when not hungry. Warm fresh milk from the mother should be put into a clean pail and held near the floor, in front of the calf, which will generally begin to "nose about the pail. Once it gets a taste of milk, it will usually drink without further trouble. Often, however, it is necessary for the attendant to put one or two fingers into the calf's mouth, drawing the hand down into the milk as the calf begins to suck the fingers. The calf in this way

(Continued on Page 26)

— Where Raleighs Rightly Reign —
FOREST HILLS FARM

F. S. REYNOLDS, Owner   DAYTON, OHIO

Is One of
America's Premier Jersey Breeding Establishments
Specializing in the

RALEIGHS
Our Herd Sires Are:

Fontaine's Raleigh 105374
Winner of 15 Championships, Gold Medals, Cups and First Prizes.
Sire of 10 in the Register of Merit.

Sire—Raleigh's Fairy Boy 83767, Grand Champion and sire of 56.

Dam—Fontaine's Gold Medal 203636, Gold Medal, public butter test, May, 1906, 3 lbs. 6½ oz. in 24 hours from 50 lbs. 10 oz. milk, 123 days after calving, 5 years old; the world's record for 5 years, and now exceeded by but 5 oz.

Noma's Perfect Raleigh 164511
A many times prize winner.

Sire—Flora's Queen's Raleigh 130251, nationally known as "The Wonder Sire," whose get was undefeated in the show ring, winning more prizes in two years than any other sire of the breed.

Dam—Fontaine's Noma 408688, one of the splendid cows at Longview Farm. She is by Perfection's Raleigh and out of Undulata Nena, a Register of Merit daughter of Fontaine's Chieftain.

Occasionally we reserve for breeding purposes a few choice young bulls out of Register of Merit dams.
If you are in search of a future herd sire it will pay you to visit
OHIO'S GREATEST HERD OF RALEIGHS AT

FOREST HILLS FARM   -   DAYTON, OHIO
J. R. BOND, Manager.

OHIO AND THE OHIO FARMER

Due to its unexcelled farming resources and the progressive character and business sagacity of its farmers, Ohio occupies a leading position in the production of farm products and in the breeding and raising of livestock, and Ohio Jersey Breeders lead the nation.

Because of the practical good derived from reading its columns, and because of its character, integrity and enterprise, The Ohio Farmer most thoroughly enjoys the patronage, and has deservedly won the confidence of the best, most up-to-date and business-like farmers of this great state.

Subscription Rates

$1.00 for one year $1.50 for two years
$2.00 for three years $3.00 for five years

Wise livestock breeders are finding it advantageous to patronize the advertising columns of "Their Own Home Farm Paper." It brings results. Write for our special rates for livestock advertising.

THE OHIO FARMER   -   Cleveland, Ohio

Prevention of Scours

By R. S. SMITH,
V. P. Ohio Jersey Cattle Club.

"An ounce of prevention is worth a pound of cure." This adage applies exactly to the blight of calf raisers—common scours.

Generally more difficulty in the raising of calves is due to scours than any other trouble.

Scours is an intestinal disease caused by the growth in the intestines of a certain kind of bacteria. The only entrance that these bacteria have to the intestines is through the mouth of the calf. This means that the calf incurs scours either in its food or drink.

Unsanitary quarters, insufficient ventilation, dark, filthy stalls, all make ideal places for the "Scours Bacteria" to breed and grow. Unclean milk pails or calf buckets and feeding boxes contribute their share.

Scours can spread very rapidly among a bunch of calves. Therefore at the first sign remove the calf to a clean pen by itself. Cut the milk feed to one-half and in this milk thoroughly mix one egg. Give an egg in three successive feeds. Then gradually raise the amount of milk feed to normal amount. This simple treatment very seldom fails if administered in time.

In five years I have never had but one mild case of scours and that was successfully checked by the above method.

Prevention is much easier than curing. Let's give our calves clean stalls with plenty of ventilation, sunlight and dry bedding. Then let's feed them clean food and let them drink out of clean utensils and our troubles with scours will almost wholly be eliminated.

OHIO HAS 135 REGISTER OF MERIT BREEDERS

Ohio Jerseys in the 50 Lbs. Test.

Month.	No. Herds.	No. Cows
September	28	51
October	39	71
November	34	64
December, 1921	50	101
January, 1922	50	102
February, 1922	37	72
March	52	123
April	63	138
May	68	188
June	60	132

The Ohio Jersey Cattle Club has pledged its word to have at least 150 herds conducting Register of Merit work before January 1, 1923.

In the United States at the present time there are 915 Register of Merit Breeders.

Results on Four Milkings Per Day

Many times I am asked how much difference it makes to milk my cows four times a day, and the table below will show what happened to five cows which we changed from four to two milkings on May 2d. The result is very interesting and significant to me and probably will be so to others with cows on official test.

Together with the reduction from four to two milkings the grain is also fed twice instead of four times. The quantity is therefore necessarily reduced, but not in half. The cow can not take as much in two feeds as she did in four, but they are all getting a total of about two-thirds as much as when on four milkings.

The drop in milk is undoubtedly due as much to the smaller amount of grain consumed as to the fewer milkings, but I lay the blame almost entirely on the fewer milkings as it is only practical to grain them at milking time as otherwise no labor would be saved and there would be no economy or saving of time in two milkings instead of four. Otherwise they are given the same care, hay and water.

Perhaps I should give my reason for changing these cows. It is obvious that they did not give enough milk to pay for the extra time and feed as they are almost all ten months or more in milk and I need all the help I can scrape together to get the place cleaned up ready to put the cows in the new barns. Dropping these cows releases another man for outside work without hiring anyone extra.

	Four milkings May 1, lbs. milk	Two milkings May 8 after one week of the change lbs. milk
Interested Dulcet	29.4	22.8
Bayleaf Sybil	30.4	25.3
Grover's Golden Gift	20.9	16.8
Hope You'll Do 2d	28.9	19.7
You'll Do's Golden Fleece	24.0	17.4

Hugh W. Bonnell, Ohio.

Long's 'Register of Merit' Jersey Stock Farm

A. M. LONG, Prop.

URBANA, OHIO

Our Herd of High Producers is rich in the blood of Gamboge's Prince, Financial King, Raleigh, Golden Maid's Prince, Noble and other Island bred sires.

Many of the females are by and bred to

Lucy's Handsome Prince 125823
Financial Bessie's Raleigh 150288
Lucy's Champion Sultan 185346.

I would like to sell

Lucy's Champion Sultan 185346

Born November 18, 1919.

Sire—Lucy's Handsome Prince 125823.
Dam—Sultana Pink 264643. She has two records to her credit, including 11,324.7 lbs. milk, 591.26 lbs. fat.

Young Stock from Register of Merit dams for sale at all times.

Write for Description and Prices.

A. M. LONG, Urbana, O.

THE OHIO EDITION
of
The National Stockman and Farmer

Has
100,329 Subscribers
In
The State of Ohio

A great many of these subscribers are dairymen and livestock growers. Just the people you should reach to sell your surplus stock.

Write for full particulars and rates.

The National Stockman and Farmer

'The World's Greatest Farm Paper'

PITTSBURGH, PENNA.

FEEDING AND MANAGING
(Continued from Page 23)

gets a taste of the milk and often begins to drink without further coaxing. If not, the process must be repeated. Sometimes, however, the calf can not be induced to drink in this way, and force has to be resorted to. In such case the feeder, facing the same direction as the calf, should straddle its neck and back the animal into a corner. The pail of milk should be held in one hand and the nose of the calf grasped with the other, two fingers being in its mouth. The nose of the calf is then forced into the milk, when it will usually begin to drink.

Sometimes a valuable calf, too weak at birth either to suck the cow or to drink from a pail, can be saved by feeding from a bottle, either with or without a nipple.

CLEANLINESS ESSENTIAL Cleanliness is absolutely essential to the successful raising of calves. This is equally necessary in feed, pens, bedding and pails or utensils. All milk fed should be fresh and clean, and the same is true of other feeds. Calf pens should always be kept clean and be filled with plenty of dry bedding. Great care should be taken in washing the milk pails. These should be thoroughly scalded with boiling water or sterilized with steam if possible. Discarded feed should be removed from the feed boxes, which should be thoroughly brushed and cleaned each day. Attention to these details is the best preventive of disease. Nearly all disorders or diseases of the calf are caused either directly or indirectly by lack of cleanliness.

Certain infections causing chronic diarrhoea or scours, either contagious or otherwise, are discussed under the heads "Scours from Indigestion" and "White Scours," but there are many small disturbances of the calf's stomach and digestive system which hinder growth and development that are caused by bacteria. Filth and dirt are the natural breeding places of bacteria. Elimination of filth usually means freedom from disease.

PASTEURIZATION OF MILK Milk from cows infected with a communicable disease, such as tuberculosis, should always be pasteurized (heated to 145 degrees Fahrenheit and held at that temperature for 30 minutes) before it is fed to calves. When separated milk from a creamery is fed it should always be pasteurized, because it is practically impossible to know that such milk is free from infection.

QUANTITY AND QUALITY OF MILK The quantity of liquid feed that a calf needs depends upon the size and age of the calf and to some extent upon the kind and condition of the feed. At birth a 50-pound calf should have eight pounds a day, while a 100-pound calf should have about 12 pounds. It is better to under-feed at the start than to over-feed. Many beginners make the mistake of
(Continued on Page 84)

My Personal Experience in Starting a Pure-bred Herd and Dairy

By L. P. BAILEY, Owner,
Belmont Stock Farm, Tacoma, Ohio.

I was not raised in a dairy section and knew very little of dairying until I was nearly thirty years old. The farmers wives all kept a few cows, not exceeding four or five. The feeling among men and boys was that milking cows was women's work. When asked to aid in the milking, I complained that it cramped my hands to milk cows. Later I realized that my wife, with her few cows, was making more money than I was making off the farm. I really had to go to her often to borrow a little expense money and to buy the equipment needed on the farm; this she readily gave me. Finally, I think she realized that she was loaning me more money than the "traffic would bear," she asked me to go in partnership with her in the cow business. I said, "Oh, it cramps my hands to milk." She said that if I would get interested in the income from the cows that that cramping hands and tired feeling would leave me. She forced me to form the partnership and to milk cows. I soon realized that the income from the cows was greater than that from the other farm products and I was really helping to make that income by milking and caring for the cows. I became interested and after that, to this day it has never cramped my hand to milk cows.

1874 to 1880 my wife had, among others, two good cows of no known breeding—just common mongrel cows. No pure-bred dairy cows in that section. Since then I have owned hundreds of pure bred cows, but among all, very few better butter producers than those two common cows.

Those two cows gave Mrs. Bailey quite a reputation as a dairy woman. We bred those cows to the best type dairy bulls—none pure-bred—we could find. Other farmers' wives were very anxious to get the heifer calves from those two cows and the men who loved their wives wanted the bull calves for breeding. They offered us three or four times as much for those calves as we could sell the calves from our other cows. We needed the money so badly that we sold them. Our neighbors appreciated getting them and we felt that we were doing ourselves an injustice in selling them.

We watched the development of those calves into cows—all disappointing—not one of the heifers ever developed into a cow equal her dam.

This experience caused me to study breeding problems. I realized that in animals and men that the character of ancestors often influence the offspring even many generations down the line.

We belong to the Caucasion race of men. The characteristic white is so fixed in our ancestry that we know to certainty that every child born to white parents will be a white child with the type and markings of the race.

If there was any Negro blood in any of our ancestry, no matter how many generations back, such parents would have an innate fear that their offspring would carry some of the markings of the Negro Race—kinky hair, flat nose or thick lips, proving that purity of blood through ancestry is most desirable.

1878 I became the owner—mainly through the pleadings of my wife—of a registered Jersey bull, cow and a heifer calf. They proved to be good ones, true to their ancestry, the offspring always equal or superior to their dams. Our interest in the Jersey cow increased; we bought as many high grades as our means would permit.

The reputation gained through our two "sport" mongrel cows aided us very much in our work with pure-breds.

1873 we bought a farm of 116 acres, carrying a debt of $4,000, paying 8 and 10 per cent interest, expecting to pay for the farm keeping sheep. Wool at that time ranged in price about sixty cents per pound. I bought good size Merino ewes, using Southdown rams, getting early mutton lambs. This, the first year, was quite profitable, but the next year the price of wool dropped to thirty, then soon to twenty cents per pound, and in a few years my fine, healthy looking lambs got the stomach worms, dwindled to almost nothing and died. This condition prevailed for a few years, we were forced out of the lamb and sheep business. Our debt increasing.

The depreciation in values of all farm products, beginning with the great money panic of 1873, had even much more to do in causing our financial troubles than did the diseased sheep.

(Continued on Page 103)

PLAY SAFE

Thousands of cows die every year from swallowing nails, slugs, staples, tag fasteners and baling wire with their feed.

These dangerous scraps of iron and steel rupture the walls of the stomach and work their way into the animal's heart.

This cannot happen if you feed

LARRO is the **only** dairy feed that is absolutely free from this death-dealing junk.

It is the only dairy feed which, in its finished state, passes over a powerful electro-magnet. Not a single particle of iron or steel can remain in LARRO.

FEED LARRO

The Larrowe Milling Co.
8047 Hamilton Ave.

DETROIT - - - MICH.

HAMER FARM JERSEYS

Combine the blood of the families that produce outstanding cows.

Our Herd Bull:
You'll Do's Theater Prince 194424

Sired by Theater Cup Prince 175372.

First dam by Rower's You'll Do 146784.

Next three dams three times to Imp. Rustic Ivy, by Oxford You'll Do 4075, with two lines to Mabel's Raleigh through Rustic Sigmond 114743.

Digest: Fauvic Prince 107961, the Peter the Great of all dairy sires.

Theater Cup Queen 382-323, the pinnacle of Jersey cows.

Imp. Rustic Ivy, one of the very greatest daughters of Imp. Oxford You'll Do, etc.

Surely up to the minute, combining extreme beauty and extreme production, and looks the part.

Twenty-five Cows and Heifers, of Raleigh, Majesty, Plymouth Lad blood lines, safe in calf to him (You'll Do's Theater Prince) for sale.

Also Young Bulls, by Nan's Jolly Raleigh 183914, junior champion at Ohio, Illinois and Tennessee State Fairs and National Dairy Show in 1921; sired by the "Wonder Sire," Flora's Queen's Raleigh 130251.

Hamer & Lockwood
HAMER FARM

LEWISTOWN - OHIO

(Logan County)

A Few Veterinary Medical Suggestions

By FRANK E. WELLS, Specialist on Jersey
Cattle—Veterinarian, Westerville, O.

THERE are many ailments which are more or less common among dairy cattle, and they so often occur under conditions, where professional services cannot be readily obtained, and the owner or attendant is compelled to render first aid and to rely upon their own resources entirely.

Some of the commoner diseases can probably be avoided by the owner or attendant through early and simple medication. So often, the farmer or dairyman resorts to the method of drenching when administering to cattle, because as a rule, the drench is usually simple, and also it is given in large quantity.

Improper methods of drenching have frequently caused the death of cattle as the result of complications following such faulty methods of administering medicine. The point, in mind, here, is that the drench for the cow should be given through the mouth and furthermore should be administered slowly, and the animal's head should by no means, be held too high. If difficulty in swallowing is noticed, cease drenching at once. A deviation from this method of administration may result in the medicine passing into the trachea, thence to the lungs, causing foreign body pneumonia, which in a large percentage of cases, means death.

The attendant, should be reasonably certain that the medicine which he is giving, has some therpeutical value, for it so often happens that cattle are given drugs, which have no curative value whatsoever, and may even cause an aggravation of the condition, which already exists.

For cases of emergency, it is well for the stockman to have a medicine cabinet of his own, containing a few simple remedies, which will suffice as first aid. The following list would, I believe, answer such a purpose:

(Magnesium Sulphate) Epsom Salts	5 pounds
Linseed Oil (raw)	1 gallon
Castor Oil	1 quart
Glycerine	1 quart
Lysol or Creolin	1 quart
Cotton	1 pound
Ginger (powdered)	1 pound
Spirits of Turpentine	1 pint
Krecamph or Formalin	1 quart
Gauze Bandages, 2½-inch	½ dozen

"In order to impress upon you the usefulness of this list, I will mention a few cases, when it is available. It so often happens that a cow needs to be purged, and the drug used depends upon the nature of the trouble and the age of the cow. If a cow has a case of indigestion, whether with or without constipation, epsom salts is usually given in one (1) to two (2) pound doses, according to the size of the animal. If, on the other hand, there is an indication of more or less inflammation in the intestinal tract, glycerine or raw linseed oil in from one (1) pint to one (1) quart doses, serve a better purpose; or, if the patient is young, castor oil or glycerine in one (1) to six (6) ounce doses should be preferred."

Camphor is also very useful, especially in combination with a base such as lard or vaseline, as a local application in the treatment of an inflammatory condition of the udder, being applied by thorough rubbing, following hot fomentation of water.

Turpentine is useful combined with some good non-irritating base, in cases where there is a great deal of flatulency.

Krecamph or Formalin in one (1) ounce doses in one (1) quart of warm water or in one (1) pint of glycerine, is a good preparation for some cases of bloating.

"A disinfectant and antiseptic, either in the form of creolin or some other coal tar product, is needed more frequently than any other medicinal agent, especially in the treatment of wounds, cow-pox and lice."

Digestive disturbances very often require something in addition to the ordinary purge, in which case a great deal of good may be obtained by giving ginger in the form of a tea, in from one (1) to two (2) pints, three (3) times a day, as a drench. As well as being a stomach tonic, ginger has a decided stimulating effect.

When surgical or antiseptic dressings are indicated, the use of sterile cotton and bandages are very important, and should always be available.

In this brief article, I have tried to encourage the adoption of a few fundamentals, which, if closely adhered to, should be of material aid, in overcoming some of the obstacles met with in dairy husbandry.

Buy no feeds that can be supplied cheaply at home. Too little thought is being given to providing this cheap home supply of grain and roughage.

CLERMONT COUNTY JERSEY CATTLE CLUB
FRANK WALMSLEY, Secretary - BRANCH HILL, OHIO
Write Me for List of Animals for Sale in Our County.

SEVEN OAKS FARM - - Loveland, Ohio
Small Select Herd of Raleighs and Nobles

BONNIE BEAUTY'S COMBINATION, Senior Herd Sire

Sired by Trouville Combination, Imp., a producing grandson of Noble of Oaklands. Dam, Bonnie Beauty of Grouville, first prize 3-year-old at National and Brockton, Mass., 1916; R. of M., 762 lbs. of 85% butter; she is the highest tested daughter of the great Maiden's Glory.

Daughters and Granddaughters of the following sires, with creditable R. of M. records or on test, are: Eminent, Golden Maid's Prince, Noble of Oaklands, Raleigh's Knight, Fontaine's Raleigh, Cute's Noble, Fern's Oxford Noble, Fairy Glen's Raleigh.

Young Bulls, sired by Bonnie Beauty's Combination, all show prospects and priced right.

Bonnie Beauty's Noble Glory, ready for heavy service. A prize winner on 1921 Circuit, out of a granddaughter of Noble of Oaklands, in R. of M. with 606 lbs. of 85% butter in 365 days.

Two unnamed bulls.

Herd under Gov't Supervision.
FOR PARTICULARS ADDRESS

LOUISE'S FAIRY LAD, Junior Herd Sire

Cut taken as a senior calf. He was third prize senior calf, Columbus and Indianapolis, in 1921, in such classes as 23. He is sired by Belle's Fairy Boy, a prize winning grandson of Fairy Glen's Raleigh. Dam, Fern Fox's Louise, ex-champion 2-year-old of Tennessee.

GEO. A. SAWYER, Union Trust Bldg., Cincinnati, Ohio

Robinson Bros. and Clark
Plain City - - Ohio

Federal Accredited Herd

Our Herd Sire:
Sybil's Gamboge Lad 163426

Sire—Sybil's Gamboge, P. 5260, H. C.

Dam—Oxford's Model Lady, by Oxford Modeller, P. 4562, C.

We have a few select sons out of Register of Merit dams to sell at reasonable prices.

WEOHONA FARM
E. H. FITCH, Owner
Hudson (Summit County) Ohio
(On the Paved Road between Cleveland and Akron)

My herd is rich in the blood of **EMINENT** and **GAMBOGE'S KNIGHT.**

Only the most select animals chosen as a foundation.

At present I have two good bull calves that I will trade for well bred heifers or sell at a reasonable price. If interested write for particulars.

WEOHONA FARM
The Home of Good Jerseys
Hudson, Ohio

How to Form a Jersey Calf Club

A JERSEY CALF CLUB is an organization of boys and girls from ten to nineteen years of age, each of whom desires to own, feed and develop a Jersey heifer calf, either grade or registered.

A constitution and by-laws are furnished for each club. This provides for the necessary officers, including an advisor, who is a man of experience and judgment, usually the town banker, county agent or leading merchant or farmer. The by-laws prescribe the manner in which the calves are to be cared for and the manner in which the bull shall be maintained.

Within a year from the date of purchase, at a suitable time of the year, an annual calf day will be observed, at which time the members will bring their calves together. This will be a day of instruction for them. The calves will be scored by a competent judge, and the points which determine a calf's place in the judging ring will be fully explained. Instruction on the feeding, the breeding and the developing of a dairy heifer will be given.

Prizes should be awarded as follows: (1) Prizes for the best calves from the judge's point of view. (2) Prizes to the members whose calves show best care and treatment.

Sales are sometimes held in connection with these meetings, but it is better to have boys and girls keep their animals than to sell them, unless compelled to.

Some prominent man or association must take the initiative in organizing clubs. The local banker is usually the man to do this, but often the agricultural leader is the logical person.

The proposition is talked about in the school, in the church, in the home and on the street. The local editor writes about it in his paper. Liberal publicity can be had and is extremely valuable. A number of boys and girls are called together and the matter is explained to them. They go out among their associates, and before long the idea has gained favor throughout the community.

Some of the most successful clubs have been organized by bankers who have inserted liberally paid-for advertisements in their local newspapers explaining the plan. These advertisements may contain a coupon for signature of the boys and girls and their parents or guardians, so that the organizer may know when a sufficient number are ready to insure a successful start.

The boys and girls are told how money will be obtained with which to buy the calves, and applications for membership are presented to them for their signatures.

The Ohio Jersey Cattle club will help in finding places where suitable calves can be bought.

Write Joe Morris, Westerville, Ohio, Secretary Ohio Jersey Cattle club.

THE PLACE OF THE BULL IN THE HERD

The real dairyman uses only high-class sires.

A bull should be more than just a bull. Ancestry counts in bulls as in men. Doubtful ancestors are an expensive luxury in the dairy business.

Two crosses to a good bull by actual test raised the yearly milk record from that of the scrub cow of 3,875 pounds to 12,804 pounds in the granddaughter and the fat from 193 pounds to 483 pounds.

A good bull may so increase the milk production of a herd and so increase the value of offspring of the cows as to amount in a few years to the value of a first-class pure-bred herd.

Having selected the best bull you can afford, do not butcher him until his worth has been ascertained. You may have a gold mine in him.

Many a dairy community could be prominently on the map of the dairy world if it would form a Bull Association and acquire a first-class community bull.

The fixed charges of maintaining a cow that milks 4,000 pounds of milk a year is but little less than those of maintaining a cow that milks 6,000 pounds of milk. A good herd sire will put the herd on the road to cheaper production by increasing production.

The average production of milk of the dairy cow in the United States is 3,527 pounds; Denmark, 5,666 pounds; Switzerland, 6,950 pounds; Netherlands, 7,585 pounds. What a possibility there is for the dairy farmer of the United States. The purebred sire of first quality is the surest start on the way to these obtainable better goals.

When a heifer, at her first freshening at less than two years, will produce more than twice as much milk as the average mature cows in the dairy herds of the country, there is no doubt that it pays to have a pure-bred bull at the head of the herd.

COME TO OHIO FOR GOOD JERSEYS — THREE THOUSAND BREEDERS WAITING TO SERVE YOU.

SUNNYSIDE FARM JERSEYS
ARE WONDERFUL PRODUCERS
And Transmit Their Inherited Re-productions to Their Progeny

In selecting foundation stock, buy from a herd that has made a record under every-day farm conditions.

Our records have all been made with ordinary farm feed, attention and care, yet our herd stood first in the Barnesville Cow Testing Association in 1920.

Buy Gold Medal Blood

Lady Blythe, the foundation of our Lady Blythe family, has three Gold Medal daughters, one granddaughter now going at the rate of a Gold Medal winner, and three young daughters who are contenders for Gold Medals in the future. "It's in the blood." All her offspring show the evidence of prepotency, production and type.

A few choice young Bulls out of high producing Register of Merit dams for sale, also a few choice Heifers from R. of M. sire and dams.

L. J. TABER - - Barnesville, Ohio

TUBERCULOSIS IN LIVE STOCK
(Continued from Page 21)

isms of tuberculosis, a large part of the milk consumed daily is not pasteurized, and some of the milk so treated is not always made entirely safe.

TUBERCULOSIS A DECEPTIVE DISEASE If tuberculosis were similar to foot-and-mouth disease in cattle, swine, and sheep, which causes rather spectacular symptoms, it would arouse immediate alarm among the live-stock owners, who would insist upon its immediate eradication; but because it is generally slow in developing and its symptoms commonly are not easily recognized from the general outward appearance of the animals, many people believe that it does comparatively little damage among live stock. Contrary to such opinions, however, the loss from tuberculosis is one of the heaviest taxes imposed upon our live-stock industry, amounting, probably, to at least $40,000,000 a year in the United States.

PREVALENCE OF TUBERCULOSIS In every State and Territory in the Union there is some tuberculosis among cattle and swine, though the degree varies considerably. In some States it probably exists quite extensively, the percentage varying from 5 to 30 per cent of the cattle population, while in certain others investigations indicate that less than 1 per cent of the total of beef and dairy cattle are tuberculous.

Tuberculosis is known to exist also quite extensively among cattle and swine in all the European countries; in fact, no part of the world is known to be free from it absolutely. There are, however, some restricted regions where its presence is not known, or it exists to but a very moderate degree.

Until cattle from the eastern part of the United States were introduced into the Middle Western, Western, and Southern States, tuberculosis among live stock in those regions was unknown so far as we know. The disease at that time was confined to the herds east of the Allegheny Mountains. It was known then that a considerable percentage of herds in those States were affected, but live-stock owners were not inclined to consider tuberculosis as of very great economic importance or dangerous to human health. Therefore very little progress was made in its eradication. As the Central and Western States became settled and cattle were moved westward the disease spread much more rapidly than is generally realized. The spreading in those areas is due, of course, to the fact that the live-stock industry occupies a more important part in agriculture than in the Eastern States. Cattle are traded in more extensively and are continually being shipped and

(Continued on Page 36)

COME TO OHIO FOR GOOD JERSEYS —— THREE THOUSAND BREEDERS WAITING TO SERVE YOU.

Correct Method of Holding a Public Sale

By TOM DEMPSEY, Ohio.

ASSUMING that you wish to dispose of your herd, or a portion thereof, the first important step is to employ a competent sales manager and then abide by his instructions. His part of the work to make a successful sale for you will cover the following:

He will see your cattle and suggest how to get them in good shape for sale day. Then he will procure all possible data from you as to their names and herd numbers, breeding dates, records, etc., so as to be able to correctly tabulate your pedigrees and print your catalog. He will prepare all advertising copy to be used in connection with your sale and at the proper time will furnish you with copy for your local papers and for posters to be put up in conspicuous places throughout your immediate neighborhood and county. The big feature of his work will be to get prospective buyers to your sale.

Now comes the part in which you should co-operate: if your cattle have an ugly, unattractive appearance about the head, get busy with a saw and trim the horns down, making them as symmetrical as possible. Then with a sharp knife, rasp or file, round the points similar to their original form. If the horns are coarse and rough, a thorough scraping with double-strength glass will soon smooth them up. Follow this by cutting a sheet of emory cloth into strips about one inch wide and work automatically as though you were polishing a shoe. It is advisable to use first a coarse grade of emory cloth and finish with a very fine one. Next apply sweet oil or vaseline to the horns and rub frequently with a strong, soft cloth or a rag of some kind. This is about the most economical manner in which to accomplish this necessary part of the fitting. With your clipping machine, go over the head carefully, clipping both the inside and outside of the ears as well as the top of the head around the base of the horns.

It is advisable to clip the udder as well as clipping the tail from the croup down to the top of the switch. Spend plenty of time brushing your cattle, always win the hair—never against it. Blanket them if at all possible to do so. The night before the sale wash the switches thoroughly and braid them, being sure your braids are tight so they will remain until the next morning. After unbraiding, give them a good combing to obtain a bushy appearance.

The general practice is to milk the cows the night before the sale but not on the morning of the sale, hence when led into the ring the udders will have a filled appearance.

Now, in the preparation of your cattle it is very desirable that you should have them accustomed to leading so, when brought into the ring, they will be easily handled. Another important feature is to be sure whoever leads the cattle into the ring is familiar with them. This attendant should be wide awake and know how to show them to the crowd to the best advantage. Many good animals have gone through the sale ring at less than their value when, if properly handled, they would have sold for more money.

It is advisable for the man most familiar with the cattle to be always at hand to answer any and all questions promptly which may be asked by intending buyers.

Don't for one moment overlook the fact that the general public appreciates condition and appearance, and every dollar you spend in feed and care will be worth two dollars to you at selling time.

In the meantime as your herd is getting into shape, prepare transfer application blanks so there will be no unnecessary delay in turning over to the purchaser the transfers on the cattle he buys. In case there are some young animals to be sold which have not already been registered, by all means get busy and have papers on hand at the time of sale. In this connection, if you write Mr. Gow, the secretary of the American Jersey Cattle Club, explaining that you are having a sale at the given date, he will rush the papers through for you.

There is no special rule about the transferring of an animal to the buyer after a sale, but we recommend that as soon as the animal is paid for you issue and sign an application for transfer of said animal, and make the purchaser an allowance sufficient to correspond with the American Jersey Cattle Club fee for issuing transfer. You will find in the end this is the best manner to handle such transactions.

About two weeks prior to the sale date your sales manager will send you about 150 catalogs of your sale, and a supply of envelopes. You should distribute about fifty of the catalogs, keeping about 100 to be used on the day of the sale.

In the average case we recommend plac-

(Continued on Page 35)

Herd Record Book

Each sheet contains ample room for every breeding and production entry.

For Sample Sheet
Send Name and Address
To

Robert L. Fleming
PEDIGREE SPECIALISTS
Canfield, Ohio

Guernsey County Jersey Cattle Club

Breeders of the Leading Blood Lines

MAJESTYS
    OXFORDS
        GAMBOGES
            NOBLES

—o—o—

Visitors Welcome
At All Times.

Make Your Wants Known to

FRANK S. CASEY,
President,
Cambridge, Ohio.

O. C. McMUNN,
Secretary,
Lore City, Ohio.

Established 1909   U. S. Accredited Herd No. 2370

W. E. RINEHART & SONS
OWL-INTEREST JERSEYS   **MAGNOLIA, OHIO**

Why Do People Say: If You Want REAL JERSEYS Go to RINEHART'S?

Rinehart's Choice Owl 167721

Why did Mr. R. H. Shriver, of Diamond, Pa., buy his foundation herd from Rinehart's after traveling six months, visiting herds from his home to the Pacific coast?

Why do you always see Rinehart's Jerseys way up in the 50-pound list?

Why does Rinehart expect four out of the first six daughters of Compound Interested Prince Owl 137849 to freshen for one Gold Medal and three Silver Medals, all under five years old?

Why is the herd clean?

BECAUSE they were bought clean, weeded, selected and bred for excellent type and know inherited production.

Mr. Shriver says: "No culls, all good, expect calves from these great cows sired by Rinehart's Choice Owl 167721, whose dam has 708 lbs. of fat and if she drops a living calf in August, 1922, will be a Gold Medal cow. His dam, both grandams will be Gold Medal Jerseys, and his sire, Sibley's Choice 83040, now a Silver Medal sire, has two Gold Medal daughters with four more on test, either going at a rate to qualify for the third Gold Medal daughter.

Consider these facts carefully, then see "Rinehart's Jerseys" before you buy elsewhere.
Young Stock for Sale at All Times.
Quality Considered, My Prices Are Exceedingly Moderate.
Write for additional information.

CORRECT METHOD OF HOLDING A PUBLIC SALE
(Continued from Page 33)

ing a three inch double column advertisement in two or three of your local or adjacent county papers for three insertions at intervals of one week apart. This will materially help to bring out the people in your own vicinity. It is natural to have a lot of spectators, and your sales manager always appreciates a large local, orderly crowd, as it does much to boost the Jersey cause. In placing your county advertisement we find that invariably said papers will gladly publish any news of interest relative to your cattle, so make it a point to see the editor and have him to do this for you.

Prior to sale day the sales manager will send you a set of printed numbers. These numbers should be pasted on the right side of the rumps of the animals, between the hip and pin bone. A home made flour paste is satisfactory for this use. These numbers are to be put on early the morning of the sale, and should correspond with the animal's number as cataloged.

The majority of Jersey sales are held under a tent. This is desirable but not absolutely necessary. A tent can be rented at a cost of approximately $25 to $35 for the occasion. Your chairs should be the folding type and can easily be procured from a local lodge hall or rented from a furniture dealer. It is a good plan to erect the tent the day preceeding the sale and if possible get your cattle accustomed to being led through the ring. The best size tent for an average sale is 30 x 50 feet. The average sale requires approximately 150 chairs or seats of some kind. These seats, by the way, do not need to be elevated, as the best practice is to have a four or five inch layer of shavings, sawdust or straw in the sale ring. This enables the people farther back to obtain a good view of the animals being sold. We recommend that you place the tent convenient to your barn to avoid having too long a walk in bringing cattle into the ring. Another thing—when the sale is once started, have enough help on hand to keep the cattle before the people, otherwise the sale will drag and the snap and pep needed for a quick, successful sale will be lost. Your sale ring should be about 16 x 18 feet, the cattle being led in one way and out another. Your auctioneer's stand can be made of 2x4's and one inch planks, 3 feet wide by 5 feet long, having same not more than 12 inches above ground. The table of your stand should be 42 inches high. You will observe in the diagram prepared to accompany this article the proper arrangement of sale ring and seating space.

A stout rope should be placed around the sale ring, fastening same securely about three feet off the ground.

You will note we suggest placing the clerk adjoining the auctioneer's stand, so he can see and properly record all sales as they take place. In this respect we suggest you get some one familiar with clerical work to handle this feature for you.

A custom seems to prevail in most neighborhoods that the Ladies' Aid Society or some other church organization furnish lunch for occasions of this kind. It is a good plan to have them do so, as it relieves you of the worry and trouble involved therein, and permits more attention to the business at hand, namely, the selling of your cattle. Although there is a practice with the larger breeders, which has proven a valuable asset to many of the sales, and that is the custom of serving a box or bag lunch gratis, supplementing this with good coffee and milk.

By no means prepare a lunch in your home on the day of the sale.

One feature of the sale which frequently is overlooked by many people is that of meeting the prospective buyer either at a certain hotel or train. This is an important part of every sale. Put yourself in the other man's place for a moment and consider that he has traveled possibly two to five hundred miles to attend your sale. Then upon arrival at the depot he is compelled to hunt you up or hire a taxicab to take him to your sale. Get the point? The pleasure is soon lost, and with a dampened ardor he attends your sale, but he is in no humor to buy because you lacked the proper spirit and failed to impress him with the fact that you are glad to have him attend your sale. You will find it a good policy not only to meet buyers at a given place, but to see that they are sent back to town at your expense after the sale is over.

Last, don't think because you have employed some one to handle your sale that you can sit down and forget it. Get busy and help pull in the buyers. It will pay.

WHY YOU SHOULD OWN JERSEYS

Jerseys mature earlier and live longer.

Jerseys are the most economical and persistent butter-fat producers.

Jerseys require less feed, less labor, less room.

Jerseys are more adaptable to extreme climates.

Jerseys are more hardy and less susceptible to disease.

Jerseys combine beauty and utility unequalled by any other dairy breed.

COME TO
Graceland Stock Farm

M. P. MURNAN, Owner

WORTHINGTON - OHIO

When in the Market for

Imported and Island Bred Jerseys

WE OFFER FOR SALE

A Splendid Yearling Bull

Born September 4, 1921.

Sire:
FERN'S PLYMOUTH NOBLE ....
W. R. Spann & Sons' famous herd sire.
{ Golden Fern's Noble. Grand Champion sire.
{ Jessie Plymouth. Gold Medal cow.

Dam:
ZULU OF VALLEY VIEW ....
On Register of Merit test, made 5177 lbs. of milk, 308 lbs. butter in 185 days.
{ Eminent's Warder Boy Sire of 2.
{ Majestic Zulu. By a grandson of Champion Flying Fox.

For quick sale will take $125.00

F.O.B. Worthington, Ohio.

Consult
RAY W. ELLIOTT
NEW CONCORD - OHIO

For

Majesty Bred Jerseys

No cow kept in my herd that will not produce 500 lbs. of fat as a mature animal.

Every animal in the herd now on Register of Merit test.

Special Offering on a "Tip Top" 6-Months-Old Bull

One noted Jersey breeder said he could win in fast company.

Sire—Lucette's Majesty. His first daughter to qualify as a 2-year-old made 340 lbs. of fat in 6 months; others now qualifying.

Dam—Raleigh's Xarama. She has two Register of Merit records to her credit; as a 5-year-old she made 693.48 lbs. of butter fat.

This is not a cheap calf. He is good enough to head a good herd.

Write for price.

Federal Accredited Herd

TUBERCULOSIS IN LIVE STOCK
(Continued from Page 32)

trailed from State to State and from farm to farm.

In some localities in the West, where dairying has developed extensively, it is now known that carload lots of cows purchased in other States have contained 50 per cent or more of tuberculous animals. Some herds of beef cattle in our western country have become contaminated with the disease by placing among them tuberculous pure-bred bulls and cows that came from diseased herds elsewhere. The importance of controlling tuberculosis and preventing its spread by the interstate movement of diseased animals was not so well recognized during the times of pioneer development as it is today. Consequently in the absence of regulations and inspection, diseased animals moved from one State to another. The shipment of cattle from Eastern and Northern States to the South, with the exception of dairy herds near the larger cities, did not commence until progress had been made in the eradication of the cattle tick. Therefore there is probably less tuberculosis among the herds of the Southern States than in any other part of the Nation. This favorable condition should be taken advantage of, for in all probability the live-stock industry will reach a high development in that area in future years. It is especially important that the herds of the Southern States be protected by permitting only tuberculosis-free animals to enter.

HOW THE DISEASE SPREADS IN A HERD The tuberculous cow is the greatest source of danger to healthy cattle. Any reacting cattle not promptly removed from the herd constitute a source of constant infection.

Tuberculous cattle, sooner or later, begin to give off the germs of the disease. These germs escape by the mouth, nose, and bowels, in the milk, and other discharges. The discharged germs are carried in the air for a time until they fall to the ground.

Animals in adjoining stalls may take in the germs in the feed they eat and thus contract the disease. Continuous water troughs in barns containing diseased cattle are a source of danger. Drinking holes containing material from infected animals are likewise dangerous.

Failure to clean and disinfect the premises occupied by the diseased cattle constitutes another source of danger. Infected milking tubes and the practice of feeding calves with raw milk from tuberculous cows are other means by which tuberculosis spreads in a herd.

LOSSES OF MEAT FOOD PRODUCTS Records kept by the Meat Inspection Division of the bureau show the great financial loss caused by tuberculosis every year. They also indicate how widespread tuberculosis in cattle and swine is in the United States, as the establishments in which the diseased animals

(Continued on Page 66)

THE OHIO JERSEY  NINETEEN TWENTY-TWO

1922--Dairy Calf Clubs of Ohio--1922

By W. H. PALMER, State Club Leader.

ONE of the best things that is being promoted in Ohio for the Dairy industry is the Boys' and Girls' Dairy Calf Club Demonstrations. It is a recognized fact that if the young people of a community become interested in any work, that they grow up to be men and women who are leaders in the work.

In the Calf Club demonstrations, the boys and girls between the ages of 10 and 18 are interested. They are organized into a club, each member securing a calf and agreeing to follow approved methods in rearing and developing it. The work is planned for at least two or three years. Starting with a calf under 12 months of age, the boy or girl works with it through calfhood and heiferhood and carries it through at least the first lactation period. Specially prepared literature is supplied, which deals with the problems of care and management arising in the various stages of development of the calf. By studying this literature and by observation, the young people learn modern methods of dairy husbandry and are able to see, from the results obtained, the advantages of dairying as a business.

Each club group holds a number of meetings (from six to twelve) during the year. At these meetings, they discuss their problems, have someone talk to them, or, judge dairy cattle.

Each year shows an increased interest in this type of Extension Work. Last year, there were 673 boys and girls enrolled. This year, there are 770 divided as follows:

450 in 1st year
198 in 2nd year
135 in 3rd year.

Jersey Clubs are organized in 21 counties.

With the County Extension Agencies greatly interested in developing the dairy industry, and realizing the value of Club Work in building for a permanent and enthusiastic interest and with the interest of the breed organizations in the work, more communities will take up this work.

Those who have seen the results of one or two years of this work are enthusiastic for it, because the boys and girls are being given an opportunity early to learn the essentials of the business through working with a calf of their own, and, through the group activities, learn to work together. Can anyone estimate the tremendous good that is sure to come from organized Dairy Calf Club Work,

Literature pertaining to this work can be had by writing to

W. H. PALMER,
State Club Leader,
The Ohio State University,
Columbus, Ohio.

GESTATION TABLE

Svd. Jan	Due Oct	Svd. Feb	Due Nov	Svd. Mar	Due Dec	Svd. Apr	Due Jan	Svd. May	Due Feb	Svd. Jun	Due Mar	Svd. July	Due Apr	Svd. Aug	Due M'y	Svd. Sept	Due Jun	Svd. Oct	Due July	Svd. Nov	Due Aug	Svd. Dec	Due Sept
1	10	1	8	1	8	1	8	1	7	1	10	1	9	1	10	1	10	1	10	1	10	1	9
2	11	2	9	2	9	2	9	2	8	2	11	2	10	2	11	2	11	2	11	2	11	2	10
3	12	3	10	3	10	3	10	3	9	3	12	3	11	3	12	3	12	3	12	3	12	3	11
4	13	4	11	4	11	4	11	4	10	4	13	4	12	4	13	4	13	4	13	4	13	4	12
5	14	5	12	5	12	5	12	5	11	5	14	5	13	5	14	5	14	5	14	5	14	5	13
6	15	6	13	6	13	6	13	6	12	6	15	6	14	6	15	6	15	6	15	6	15	6	14
7	16	7	14	7	14	7	14	7	13	7	16	7	15	7	16	7	16	7	16	7	16	7	15
8	17	8	15	8	15	8	15	8	14	8	17	8	16	8	17	8	17	8	17	8	17	8	16
9	18	9	16	9	16	9	16	9	15	9	18	9	17	9	18	9	18	9	18	9	18	9	17
10	19	10	17	10	17	10	17	10	16	10	19	10	18	10	19	10	19	10	19	10	19	10	18
11	20	11	18	11	18	11	18	11	17	11	20	11	19	11	20	11	20	11	20	11	20	11	19
12	21	12	19	12	19	12	19	12	18	12	21	12	20	12	21	12	21	12	21	12	21	12	20
13	22	13	20	13	20	13	20	13	19	13	22	13	21	13	22	13	22	13	22	13	22	13	21
14	23	14	21	14	21	14	21	14	20	14	23	14	22	14	23	14	23	14	23	14	23	14	22
15	24	15	22	15	22	15	22	15	21	15	24	15	23	15	24	15	24	15	24	15	24	15	23
16	25	16	23	16	23	16	23	16	22	16	25	16	24	16	25	16	25	16	25	16	25	16	24
17	26	17	24	17	24	17	24	17	23	17	26	17	25	17	26	17	26	17	26	17	26	17	25
18	27	18	25	18	25	18	25	18	24	18	27	18	26	18	27	18	27	18	27	18	27	18	26
19	28	19	26	19	26	19	26	19	25	19	28	19	27	19	28	19	28	19	28	19	28	19	27
20	29	20	27	20	27	20	27	20	26	20	29	20	28	20	29	20	29	20	29	20	29	20	28
21	30	21	28	21	28	21	28	21	27	21	30	21	29	21	30	21	30	21	30	21	30	21	29
22	31		Dec	22	29	22	29	22	28	22	31	22	30	22	31	22	July	22	31	22	31	22	30
	Nov	22	1	23	30	23	30		Mar		Apr		May		Jun	22	1		Aug		Sep		Oct
23	1	23	2	24	31	24	31	23	1	23	1	23	1	23	1	23	2	23	1	23	1	23	1
24	2	24	3		Jan		Feb	24	2	24	2	24	2	24	2	24	3	24	2	24	2	24	2
25	3	25	4	25	1	25	1	25	3	25	3	25	3	25	3	25	4	25	3	25	3	25	3
26	4	26	5	26	2	26	2	26	4	26	4	26	4	26	4	26	5	26	4	26	4	26	4
27	5	27	6	27	3	27	3	27	5	27	5	27	5	27	5	27	6	27	5	27	5	27	5
28	6	28	7	28	4	28	4	28	6	28	6	28	6	28	6	28	7	28	6	28	6	28	6
29	7	29	8	29	5	29	5	29	7	29	7	29	7	29	7	29	8	29	7	29	7	29	7
30	8			30	6	30	6	30	8	30	8	30	8	30	8	30	9	30	8	30	8	30	8
31	9			31	7			31	9			31	9	31	9			31	9			31	9

No Abortion   No Tuberculosis

Long Meadow Herd

of

Sophie's Tormentors

Every cow has a Register of Merit record, or else on test.
The average R. of M. records last year were: 9873 lbs. milk, 551.4 lbs. butter fat.

Young Stock for Sale.

M. C. EBRIGHT

Shreve, Ohio

(Wayne County)

POLLED JERSEYS

◆——◆

We believe that the verdict of the breeders of the future will be that the **Polled Character** has been Ohio's greatest contribution to the Jersey breed.

One of our aims is to supply breeders of Jerseys with **Pure Polled Sires** from Register of Merit ancestry.

Sires whose calves will all be without horns and whose daughters will be superior to their dams.

If you are interested we should be pleased to have you visit us and see our cattle.

◆——◆

Charles S. Hatfield

Box H, R. D. No. 4,
SPRINGFIELD, OHIO

Raleighs
AND
Nobles

A Few Choice Open and Bred Heifers
For Sale
at
All Times

—o—o—

J. B. HICKMAN

MARIETTA,
OHIO

A. J. Nickols

Berlin Heights - - Ohio

Breeder of
HIGH CLASS JERSEYS
My Herd Sires are:
OSWALD'S SULTAN 177475;
HILLTOP'S HAPPY DAYS 136444,
sire of 6 in R. of M.
Both are grandsons of
Golden Fern's Noble.
Only 35 Animals in the Herd,
but every one is a good one.
For Noble Blood see Me

Creek Farm Jerseys Lead in Ohio

C. C. CREEK

Owner of many high class producing Jerseys, including
FOXHALL'S LUCIA 307323
The long distance champion over all breeds in Ohio, 70,407 lbs. of milk, 3169.62 lbs. butter fat in 6 years' test.
My Herd Sire is
Creek Farm Sophie Tormentor
A son of Pogis 99th of Hood Farm, Champion A. J. C. C. Gold Medal bull.
Herd on Register of Merit test for seven years and Fully Accredited by the Federal Government.

Montpelier - - **Ohio**

THE OHIO JERSEY — NINETEEN TWENTY-TWO

Grand Champion Males and Females at Previous National Dairy Shows

1906	Rachel Benton (Cow), Hunter & Smith, Lincoln, Neb. Emanon 52299 (Bull), Hunter & Smith.
1907	Brookhill Fox 65303 (Bull), Overton Hall Farm, Nashville, Tenn. Golden Fern's Sensation 173201 (Cow), Lewisiana Farm, Fredericksburg, Va.
1908	Royal Majesty 79313 (Bull), G. G. Council. Jolly Lady of Beechwood 213915 (Cow), G. G. Council.
1909	Derry's Jolly Lad 80538 (Bull), J. F. Boyd, Rushville, Ind. Majesty's Oxford Lass 213940 (Cow) G. G. Council, Vandalia, Ill.
1910	Raleigh's Fairy Boy 83767 (Bull), C. I. Hudson, East Norwich, L. I. Bosnian's Anna 231557 (Cow), C. I. Hudson.
1911	Sultana's Golden Jolly 86180 (Bull), T. S. Cooper & Sons, Coopersburg, Pa. Great Scot's Champion 203703 (Cow), Ed C. Lasater, Falfurrias, Texas.
1912	Fontaine's Chiefton 97158 (Bull), Undulata Farm, Shelbyville, Ky. Gamboge's Tiddledywinks 252156 (Cow), White Horse Farms, Paoli, Pa.
1913	Noble's Eminent Lad 113642 (Bull), Ed C. Lasater. Ula of Fair Acres 292754 (Cow), John B. Stump, Monmouth, Ore.
1914	Allen Dale's Raleigh 109356 (Bull), Allen Dale Farm, Shelbyville, Ky. Noble's Jolly Norah 293938 (Cow), Elmendorf Farm, Lexington, Ky.
1916	Golden Fern's Noble 145762 (Bull), Wm. Ross Proctor, Barryville, Ill. Gloria Benedictine 246997 (Cow), A. V. Barnes, New Canaan, Conn.
1917	Gamboge's Vellum's Majesty 123063 (Bull), N. D. Munn, Forest Lake, Minn. Oxford Majesty's Gipsy 344076 (Cow), L. V. Walkley, Southington, Conn.
1918	Raleigh's Oxford Prince 123167 (Bull), Ed C. Lasater. Oxford Majesty's Gipsy 344076 (Cow), L. V. Walkley.
1919	Leda's Raleigh 137009 (Bull), Wm. Ross Proctor. Constance of Falfurrias 332639 (Cow), Ed C. Lasater.
1920	Fashionable Fern Lad 163968 (Bull), Longview Farm, Lee's Summit, Mo. Sly Puss P. 15051, C. (Cow), Longview Farm.
1921	Fashionable Fern Lad 163968 (Bull), Longview Farm. Brampton Seaside Lass 4671 C. J. C. C. (Cow), John Pringle, London, Ont.

Record Animals of the Jersey Breed

RECORD PRICED BULL — Sybil's Gamboge 176463, Imp. Sold in auction for $65,000.00. Owner, L. W. Walkley and Associates, Southington, Conn.

RECORD PRICED AGED COW — Gamboge Oxford Gem P. S. 21724 H. C. Imp. Sold at auction for $18,000.00. Owner, J. C. Baldwin, Jr., Mt. Kisco, N. Y.

RECORD PRICED YOUNG COW — Fern's Oxford Triumph 477786, Imp. Sold as a two-year-old for $15,000.00. Owned by Ayer and McKinney.

RECORD BULLS WITH 1000-POUND BUTTER DAUGHTERS — The Imported Jap 75265: Five daughters with 1000-lb. records. For twelve years herd sire at Meridale Farms.

Pogis 99th of Hood Farm. With five daughters with 1000-lb. records. Owned by Hood Farm, Lowell, Mass.

RECORD MILK COW OF THE BREED — Fauvic's Star 313018; 20,616 lbs. milk, 1,005.9 lbs. fat at 6 years and 11 months. Owner, A. V. Barnes, New Canaan, Conn.

RECORD BUTTER FAT COW — Lad's Iota 350672; 18,632 lbs. milk, 1,048.07 lbs. butter fat at 5 years 6 months old. Owner, S. J. McKee, Oregon.

COME TO OHIO FOR GOOD JERSEYS —— THREE THOUSAND BREEDERS WAITING TO SERVE YOU.

MARYVALE FARMS

Sybil's Gamboge Jerseys

The leading sires of the breed are represented in this herd of cows. Imported Prince Sybil 174664 stands at the head of one of Ohio's foremost breeding nurseries. His calves are fulfilling every expectation of a great sire.

COME TO OHIO FOR GOOD JERSEYS

Visitors Welcome   Correspondence Invited

—::—

Maryvale Farms - - - Youngstown, Ohio

R. C. MILLIKIN, Manager

OAK GROVE FARM

Breeders of

SOPHIE TORMENTOR JERSEYS

Experienced Jersey breeders recognize in the Sophie's Tormentor family of Jersey Cattle the acme of perfection in both breeding and individuality, but do you as a layman breeder realize that

The Sophie Tormentor Family Has Produced

The World's Champion Sire, sire of 36 daughters, each making over 600 lbs. of fat.

The Champion A. J. C. C. Gold Medal Sire, having ten daughters who average 857 lbs. of fat.

The World's Champion Long Distance Producer, 10 years test, 123,103 lbs. of milk, 7,038 lbs. fat.

The only sire of the breed, sire of ten daughters each making over 800 lbs. fat.

Twenty-three and one-half percent of all Gold Medal sires.
Thirty-three and one-third percent of all Silver Medal sires,
and many other cherished awards.

My Herd Sires are:

SOPHIE'S PREMIER
Sire of 25 in the Register of Merit.
Sire—Pogis 99th of Hood Farm, the "Champion R. of M. sire."

SOPHIE'S TORONO 23rd
His daughters are just coming in milk and give great promise.
Sire—Sophie's Torono, one of the great bulls bred at Hood Farm.

Buy a Sophie Tormentor Bull out of a Register of Merit cow from me if you want "The Blood of the World's Champion Jerseys."

Quality Considered
My Prices are Reasonable

OAK GROVE FARM

RANDALL H. ANDERSON - WEST AUSTINTOWN, OHIO

MAKING BUTTER ON THE FARM
(Continued from Page 15)

used to some extent, in the belief that it aids creaming, but investigations have shown that the loss of butter fat is as great as, or greater than, in the shallow-pan method. There is the further objection that a watery flavor is imparted to the cream and the usefulness of the skimmilk is limited, mixtures of water and skimmilk being undesirable either for household use or for calf feeding. The water dilution method therefore is not advisable under any conditions.

A centrifugal separator gives by far the best results, because the separation is accomplished in a few minutes, while the milk is still warm. The skimmilk usually contains only a trace of butter fat and is available for use at once, while perfectly fresh. Because of the ability of the mechanical separator to skim clean, it is a profitable investment unless the quantity of milk is very small.

THE USE OF A CREAM SEPARATOR A cream separator should be placed in the dairy house or dairy room where there are no odors to contaminate the milk and cream during separation. It must be set level and firmly fastened on a solid foundation so as to be rigid when in operation. If that is not done the running of the machine will cause the frame to vibrate, and as a result the bowl will wabble the bearings wear quickly, and the separation of cream from the milk will not be complete; that is, butter fat will be lost in the skimmilk. When setting up the separator a spirit level should be used to insure that the upper surface of the bowl casing is level. If the machine is set upon wood, lag screws may be used to fasten it in place; if upon cement, a bolt should be set in the floor, exposing thread enough to extend through the frame of the machine and accomodate a nut. Bolts may be set in the cement when the floor is laid, or holes may be drilled, the bolts inserted, and molten lead poured around them until flush with the floor. When cold the lead will have shrunk and should be pounded in tight. Washers or other pieces of metal may be used at the bolts to make the machine level. An especially sanitary setting for a separator may be made by setting the machine upon pieces of ¾-inch pipe about 1½ inches long. The machine is then supported upon four short posts, which makes cleaning the floor much easier.

A cream separator should be run according to the directions furnished by the manufacturer. Bearings and gearings should be kept clean, free from grit and well lubricated with good oil. Special care should be used to run the machine at the speed recommended by the manufacturer. If a speed indicator it not used, the revolutions of the crank should be timed by a watch or a clock. In turning, even pressure should be maintained on the handle throughout the revolution, as jerking causes unequal wear on the bearings and gears. The cream separator is probably the most delicate machine in general use on the farm, and should be handled with the care that its construction demands.

A separator does its best work only when run under proper conditions. It will not skim clean when (1) it is run too slowly, (2) the bowl wabbles or vibrates, (3) the milk is too cold, 90 degrees Fahrenheit being the minimum temperature for the best work, (4) the bowl parts are bent, dirty or not properly assembled, (5) particles of foreign matter get into the bowl and partially obstruct the cream outlet, or (6) the milk is nearly sour. During the winter, in order to warm the bowl, some warm water should be run through the separator so that the first milk that enters will not be cooled below 90 degrees Fahrenheit. When through separating, a small quantity of skimmilk or warm water should always be used to flush the bowl in order that no cream may be wasted.

Like all other milk utensils, the separator should be cleaned thoroughly immediately after each time it is used. Merely flushing the bowl with warm water after use and taking it apart for washing but once a day is a filthy practice and must be condemned. All parts of the separator bowl, together with the other tinware, should first be rinsed with lukewarm water, then thoroughly scrubbed with a brush in warm water in which washing powder has been dissolved. Soap or soap powder are liable to leave a soapy film on the utensils and should not be used. Soda ash or one of the commercial dairy cleansing powders is satisfactory, as either is easily rinsed off. The utensils should then be sterilized by means of the farm sterilizer or boiled for five minutes. The use of a dish towel or cloth for drying is not necessary or desirable, because the hot utensils will dry themselves, and in order that they may remain sterile they should be handled or touched as little as possible.

The thorough cleaning and sterilizing of all dairy utensils is essential to the production of butter of good flavor. Unclean utensils harbor bacteria that, when the utensils are used again, contaminate the milk and cream and develop bad flavors and thus injure the butter.

Thin cream has the same objectionable features for churning that whole milk has, though in a less degree. For that reason the cream separator should be regulated to

(Continued on Page 43)

A Pure-Bred Feed for Pure-Bred Cows

When better cows are bred—
We will make better feed for them.

UNION GRAINS

24% Protein   5% Fat   10% Fiber

The First Dairy Feed Made—
The Standard of Quality for 20 Years

—::—

Milk Record Cards Free.   Write to

The UBIKO MILLING CO. - Cincinnati, O.

FOR ABORTION AND FAILURE TO CLEAN

USE

Hood Farm Breeding Powder

Cows should always be treated with Hood Farm Breeding Powder immediately after aborting. It destroys germs in the internal genitals, and thoroughly disinfects the organs, establishes normal healthy conditions, preventing barrenness and aiding greatly in avoiding another abortion.

When cows do not clean, the afterbirth should not be forcibly removed. Inject Hood Farm Antiseptic Breeding Powder, which prevents putrefaction and bad odor, averts the danger of blood poisoning, and the placenta comes away of itself without violence or any bad effects.

Read the following:

"Shepherdstown, W. Va.
"Gentlemen:—

"I have tried several different kinds of remedies for cleaning cows up after calving and failure to clean, but can't get anything I like as well as Hood Farm Breeding Powder. For three years I had a lot of abortion, but have gotten it cleaned up now, and I think your Breeding Powder is entitled to the most credit for getting rid of the abortion. I use it regularly on every cow that calves until she is bred.
"Very truly,
"Charles S. Billinger."

Prices of Hood Farm Breeding Powder, prepaid, $1.15, $2.75 and $5.00.

Price of Hood Farm Flexible Injection Tube, by mail, 90c., or with a $5.00 order 75c.

Mail all orders for Hood's Farm Remedies direct to

C. I. HOOD CO., Lowell, Mass.

"Good to the last drop"

Mr. Jersey Breeder:
Please read what Mr. Chas. W. Bogert, Jr., of East Stroudsburg, Pa., says about Blatchford's Calf Meal.

"East Stroudsburg, Pa., Aug. 5, '22.
"Blatchford Calf Meal Co., Waukegan, Ill.
"Dear Sirs:—I have a registered Jersey calf, seven months old which I have raised on your meal. To date I have fed 200 lbs., which I purchased from J. M. Wyckoff, miller, of this town. I am going to exhibit this calf at the Monroe County Agricultural Fair from Sept. 4th to 8th, and would be glad at the same time to advertise your meal. In the estimation of cattle raisers and farmers in this community this calf is a fine specimen. If you so desire I will be glad to send you a picture of Prince's Fawn Belle, which you may use in advertising. Yours very truly,
"Charles W. Bogert, Jr."

If you have any doubt as to the value of this product in the growing of pure bred Jersey calves, write direct to Mr. Bogert or to the Blatchford Calf Meal Company.

Blatchford's Calf Meal is a proven substitute for milk; 300 pounds of milk and 200 pounds of Blatchford's Calf Meal will grow as good a calf as 2000 pounds of milk. Let us prove it to you.

Write us.   Dealers everywhere.

BLATCHFORD CALF MEAL COMPANY
Waukegan, Illinois

MAKING BUTTER ON THE FARM
(Continued from Page 41)

deliver cream testing about 30 per cent butter fat, or so rich that one gallon will yield about 3 lbs. butter.

COOLING THE CREAM After separation, the cream should be placed immediately in cold water and stirred occasionally from the bottom with a stirring rod until the temperature is below 60 degrees Fahrenheit at least, and preferably below 50 degrees Fahrenheit. Fresh cream should never be mixed with cream from previous skimmings until it has been thoroughly cooled, as the addition of warm cream raises the temperature of the older cream and hastens souring. Water is a much better cooling agent than air, because it is a better conductor of heat and is capable of absorbing greater quantities of heat. In cooling, the best results are obtained when ice water is used. A dairy farmer in a section where natural ice is produced should have an ice house and should fill it each winter. If well water alone is used, it is necessary to change it several times a day. For that reason the cooling tank should be between the well and the stock tank, so that all water pumped for the stock passes through it. A spring or a stream of cold water is very satisfactory, because it performs the work continuously, without attention.

Cooling tanks of various types may be obtained from dairy supply houses or may be made on the farm. A satisfactory wooden tank may be made of 2-inch planed cypress boards properly bolted together, painted on the outside and oiled on the inside. Concrete makes a most serviceable tank which can be constructed by any one accustomed to working with that material. A very simble and cheap cooling tank may be made also from two or more vinegar barrels—one for each cream can. Whatever the style of tank, the pipe conveying the water to it should be large enough to carry the full stream from the pump. Upon entering, the inlet pipe should be carried to within a few inches of the bottom by means of an ell and a short piece of pipe, so that the cold water may be conducted to the bottom, thus forcing the warmer water at the top through the outlet pipe. The outlet pipe should be at the end of the tank opposite the inlet pipe, of slightly larger diameter, and so high that the water will be nearly at the tops of the cans. Lock nuts and sheet packing may be used to make tight joints where pipes enter and leave the tank. Cream cans should stand on cleats in the tank, so that water may circulate under as well as around them.

In order to afford protection from the heat, a cooling tank should have a tight cover and be placed in the dairy house or under a shed, where it will be protected from the hot winds and direct rays of the sun. If water does not flow continuously through the tank it may be advisable to insulate the tank, as an insulated tank uses less ice and requires less frequent changing of water than an ordinary one. Tanks of that type may be purchased at a reasonable cost, or the insulation may be put on at home. To insulate a tank at the lowest possible cost, 6 inches of dry excelsior, shavings, or sawdust, tightly packed on the sides, bottom and cover of the tank, will serve the purpose if kept dry.

RIPENING THE CREAM On many farms it is customary to churn only two or three times a week. Where this is the case the cream from each separation should be kept in the cooling tank until about twelve hours before churning. In order that the cream may ripen uniformly, it should be placed in one receptacle, thoroughly mixed, and warmed slowly to a temperature of from 65 to 75 degrees Fahrenheit. Frequent stirrings with the stirring rod and the use of a thermometer are necessary to insure uniform and proper temperature throughout. Fresh cream should not be added after ripening has begun. The cream should be allowed to stand at the ripening temperature (from 65 degrees to 75 degrees Fahrenheit) until it thickens, assumes a glossy appearance, and is mildly sour, when it should be cooled quickly to churning temperature or below. The churning temperature is usually from 52 degrees to 60 degrees Fahrenheit in the summer and 58 to 66 degrees Fahrenheit in the winter. This cooling may be done if the cream is in a can by placing it in the cooling tank and stirring it occasionally. Ice or cold water should never be put into the cream. In order that the butter may have the desired firmness of body, the cream should be held at churning temperature or slightly below for at least two hours before it is churned. Even after it is cooled the cream will continue to sour somewhat, but when ready for churning it should still be only mildly sour, not to exceed 0.6 per cent acidity, as determined by the acidity test.

Special care should be taken to prevent the cream from becoming too sour, which has two harmful results — it gives the butter a sour, overripe cream flavor and injures its keeping properties.

The souring of cream is caused by the growth of bacteria, which are a simple form of plant life. Some bacteria produce lactic acid, and as a by-product, the flavors that are desirable in butter. Many other

(Continued on Page 44)

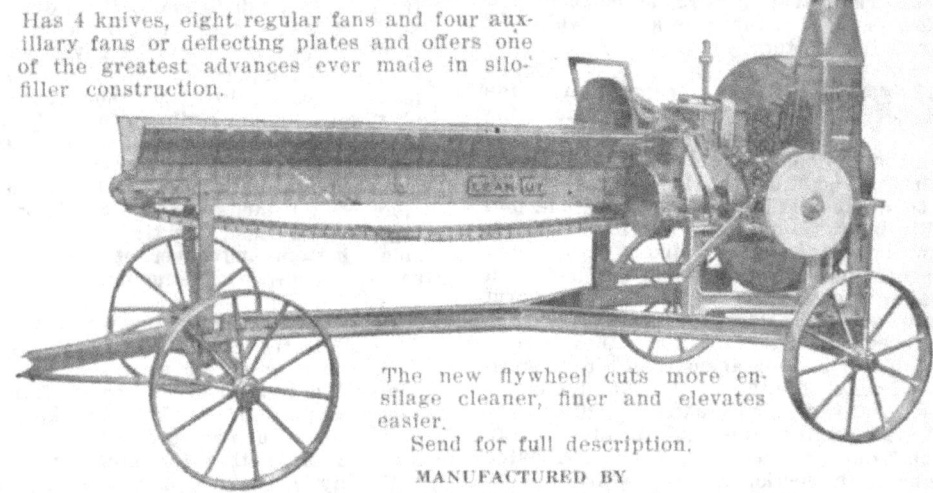

MAKING BUTTER ON THE FARM
(Continued from Page 43)

types of bacteria, however, grow and produce bad flavors at the temperature used for ripening cream. If the milk or cream has been contaminated by unclean methods during milking or by utensils that have not been properly cleaned and sterilized, "off flavors" will develop in the cream during ripening and will be retained in the butter. Undesirable flavors may be developed even in clean cream if the ripening temperature is too high or too low or if the cream becomes overripe; in fact, an overripe cream flavor is one of the most common defects in farm butter.

The organisms that develop the desirable lactic acid and its attendant flavors in the cream are very susceptible to the influence of temperature. Although they grow and produce acid in a very wide range of temperature, the flavors that are desired in butter are produced only within a very narrow range. It is therefore very essential to use an accurate thermometer and to control the ripening temperature carefully. Lactic-acid bacteria are more active in summer than in winter, and for that reason, together with the fact that the temperature of the cream during ripening is usually affected somewhat by the atmospheric temperature it is well to begin the ripening process at a higher temperature in winter than in summer. Experience will demonstrate just how to handle the cream so that is will be in the proper condition when it is desired to churn.

STARTERS In creameries it is customary to control to some extent the ripening of cream by means of "starters," which are pure cultures of lactic-acid-producing bacteria grown in pasteurized milk. The making of starters is technical work that should not be undertaken unless butter is made on a commercial scale. If the milk and cream are produced under proper conditions, there is no need for using starters. If handled under those conditions and protected from contamination, cream will develop the desired flavor when allowed to ripen or sour naturally at the proper temperatures.

When butter is made on a commercial scale, it may be advisable to control the ripening and thus make a product that is more uniform from week to week.

Commercial cultures for starter making may be obtained from culture manufacturers and from dairy supply houses. Directions for using accompany each package and should be followed carefully.

A natural or home-made starter may be made as follows:

1. Clean thoroughly and boil for five minutes three pint fruit jars and tops. After boiling, keep the jars covered to prevent the entrance of bacteria.
2. Take a pint sample of milk freshly

(Continued on Page 99)

Ohio Runs Close Second to Mississippi Winning State

By JOE MORRIS, Secretary

AFTER several years of floundering around in a sea of inactivity, the breeders of Ohio were, as if by magic, awakened to the fact that "their light was being hid under a bushel," and realizing that 'twas best to "make hay while the sun shone," many of the breeders then, as some few now, still fail to grasp the full significance of "E Pluribus Unum," nevertheless, as is an undisputable fact, that in every civic body or group there are one or more leaders, who do not fear adverse or unjust criticism, and who for the sake of benefiting the Jersey breed as a whole, are willing to sacrifice unlimited quantities of their time and ability for the advancement of their fellow breeders.

There assembled from all sections of Ohio at Townsend Hall, Ohio State University on February 3, 1921, about 100 breeders, whose purpose and aim was to reorganize the Ohio Jersey Cattle Club, and to do some real good for the Jersey cause. After a few opening remarks by Mr. L. P. Bailey, one of the oldest breeders in the state, impressed those present with the necessity of making a certed effort to place Ohio in its proper sphere before the breeders of the entire country. So ere the day was over, a new set of officers and directors were appointed and given carte blanc to proceed, if their success is eminent, it has been mostly through the guiding hand of its ambitious and energetic president, Mr. Hugh W. Bonnell, of Youngstown, Ohio. Throughout the entire year he has been unfailing in his devotion to the work of directing the policy of the Ohio Jersey Cattle Club, also considerable credit is due Mr. Tom Dempsey who financed Ohio's pro rata share of the expense incurred in obtaining the service of Mr. R. D. Canan as Tri-States fieldman.

In a brief and concise manner are enumerated below some of the activities of the club for the year of 1921:

Club reorganized February 3, 1921.

An invitation sent to 2,974 Jersey breeders in Ohio to co-operate with us to boost Ohio Jerseys and its breeders.

Membership campaign carried on by the officials of the club. Mr. Canan visited practically every section of the state. As the membership passed the 100 mark, the club began its publicity campaign through THE JERSEY BULLETIN, and other local papers.

Every member was invited to send pictures or cuts of his record animals so that he could participate gratis in the space contracted for in THE JERSEY BULLETIN. The club members were called upon and responded loyally in giving their aid and moral support to helping the Kryder ice cream bill through the Ohio legislature, said bill in brief was to standardize the ingredients, butter fat in particular, used in the manufacture of ice cream sold in our state. This bill is now in force and greatly helps every breeder of dairy cattle in Ohio.

An entire week was set aside and a series of meetings and tours were held in the state, with Prof. Erf and Mr. Canan as leading speakers, thus helping to cement a closer friendship between the state and the local clubs and the breeders at large.

If called upon the Ohio Jersey Cattle Club agreed to assist financially to the extent of $200.00 the work of the Ohio Dairymens' Association, said funds apparently not needed, as it was never called for and has been used otherwise.

A policy originally outlined was to award a silver cup to the owner of the highest testing Jersey cow finishing her record in the calender year of 1921. This was later modified and suitable ribbon will be awarded to the owner of each of the highest testing cows in the eight classes, for the same period, said awards are to be made by Mr. M. D. Munn, president of the American Jersey Cattle Club, who will present them at the annual meeting of the Ohio Jersey Cattle Club to be held in Townsend Hall at the Ohio State University, Wednesday, February 1, 1922.

An attractive booth was erected at the entrance of the Jersey cattle quarters during the Ohio State fair, at which time aside from Mr. Canan being on hand, one or more club mempers took charge, and rendered valuable aid to the many exhibitors from outside of the state as well as our own people. A banquet was held at one of the leading hotels on the night preceeding the Jersey judging. It was well attended and did much to further unite the breeders.

The club published its first issue of The

(Continued on Page 46)

W. O. Phillips & Sons
CENTERBURG - - OHIO
Breeders of High Class
REGISTER OF MERIT JERSEYS
Our Herd Sire:
Lilac's Fair Play

His first four daughters as 2-year-olds averaged 7747 lbs. of milk, 446 lbs. of butter fat.

His dam, Jersey of More Fox, is a Gold Medal cow in the Register of Merit with a record of 12,293.1 lbs. of milk, 794.90 lbs. of butter fat.

We have ten cows with Register of Merit records that averaged 520 lbs. of butter fat in 1 year.

CHOICE YOUNG STOCK FOR SALE AT ALL TIMES

Quality Considered Our Prices Are Very Reasonable

Everything tuberculin tested and herd absolutely clean. We hope soon to get our Accredited Herd Certificate.

J. H. LEWIS   J. KENNETH LEWIS
J. H. LEWIS, Jr.

Sunny View Farm

CARROLLTON, OHIO, R. D. 6
J. H. LEWIS & SONS, Props.

Registered Jersey Cattle

Herd Sire: WANDA'S GOLDDUST ADVANCE 182606, show 5 times, winning 5 blues.

Delaine Merino Sheep
Duroc Jersey Swine

BRONZE TURKEYS
Golddust Strain, the leading strain of the nation. 40 years breeding Bronze Turkeys we have produced the dream in fancy bronze color.

GOLDEN and SILVER LACED ROCKS
The new Beauty Breed of Rocks. Nothing like them.

WHITE EMBDEN AND PAN-AMERICAN GEESE

FANCY CRESTED DUCKS
All colors.

OHIO RUNS CLOSE SECOND
(Continued from Page 45)

Ohio Jersey, a booklet devoted to advertising Ohio Jerseys and their breeders. Approximately 4,000 copies were distributed, of which some 2,200 requests came from direct advertising in THE JERSEY BULLETIN. In publishing this booklet all Ohio breeders were invited to participate and share the expense by using a given amount of space. A sufficient number responded and enabled the club to publish the booklet, costing about $1,000 yet less than $200 came from the club treasury for this work. Copies were mailed upon request to several foreign countries as well as almost every state in the union. Of this booklet the Breeders' Gazette stated: "It is the best piece of breed literature we have seen," and of it THE JERSEY BULLETIN wrote as follows:

"The Ohio Jersey Cattle Club has issued its first publication under the title of 'The Ohio Jersey.' It is in magazine form, and while the original announcement was that is was to be a directory, as it came from the press it is more than a directory — it is a real publication, chock full of short articles by breeders of Ohio and elsewhere, and of course all of it corking good. The advertising is magnificent support to the secretary, on whose shoulders most of the editorial and advertising work fell. While we understand it is an annual publication, we want to put ourself on record as endorsing the appearance of such efforts in large numbers.'

Approximately twenty-five full pages of advertising appeared during the year in THE JERSEY BULLETIN, and on two specific occasions, a list showing the names and addresses of every member in the Ohio Jersey Cattle Club was published. At another time a list of local or county clubs, on another occasion considerable publicity was given in local papers to the Ohio breeders who owned state class champions.

Throughout the year, all enquiries for Ohio Jerseys were sent to the members of the association, giving them first opportunity to supply the demand.

Today there are thirty local Jersey county clubs (twice as many as any other state), and everyone of them is actually functioning on behalf of the Jersey breed. Our aim for 1922 is to increase this to fifty counties.

There are 243 paid members in the club and the entire Jersey public has learned that the Ohio Jersey Cattle Club is the strongest and largest state Jersey cattle club in existence, that it has set the pace and publicly challenges any state Jersey club to equal its record, and last but not least, that "Ohio is the best place to buy Jerseys."

The foregoing is a brief outline of the work done by the officers and directors of the Ohio Jersey Cattle Club. The bare fact that we failed to win the 1921 Jersey Bulletin Accomplishment Cup is no criterion that it will not be awarded to Ohio for 1922.—Editor.

COME TO OHIO FOR GOOD JERSEYS —— THREE THOUSAND BREEDERS WAITING TO SERVE YOU.

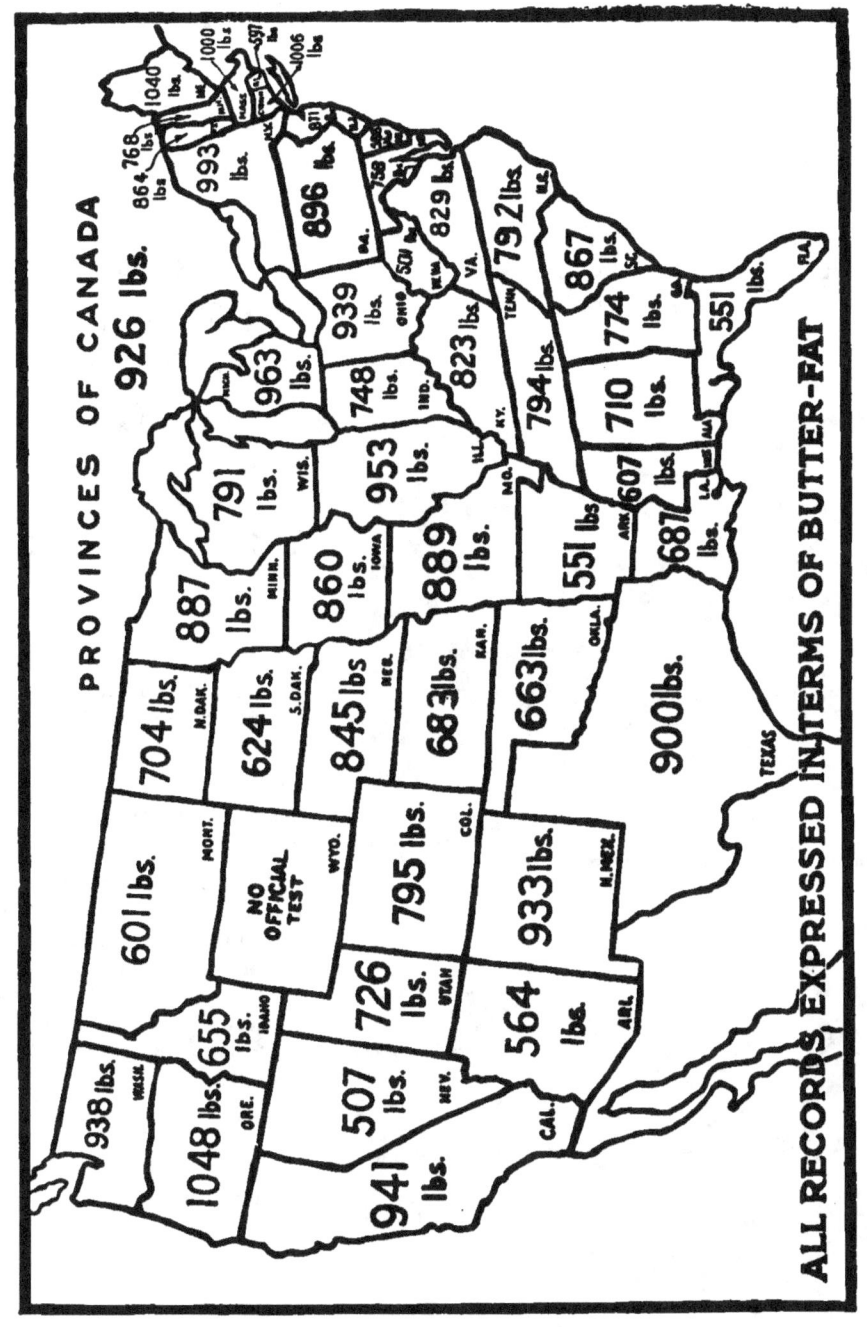

PRIDE'S OLGA OF LONGVIEW 199691
Ex-Champion Jersey of Ohio
Official test, 14233.7 lbs. milk, 908.58 lbs. of 85% butter.

ELM SHADE FARM

C. E. HARRIS & SON
Owners

Breeders of
REGISTERED JERSEY CATTLE

R. F. D. 2.
Medina - Ohio

PRIDE'S OLGA OF ELM SHADE FARM 254341
Now on R. of M. test, produced 9591 lbs. milk, 500.5 lbs. butter fat in 180 days. Daughter of Pride's Olga of Longview.

These pictures illustrate the quality of Jerseys bred and tested by us.

They are rich in the "Olga" blood, large size, splendid dairy type and real producers.

We offer at this time two good sons of Lass 89th of Hood Farms Son 165860, one of them a 2-year-old out of Pride's Olga of Longview, the other a yearling out of Pride's Olga of Elm Shade.

Other Choice Young Stock For Sale at All Times

TORONO'S OLGA LASS 455593
Now on R. of M. test, 5036 lbs. milk, 281.3 lbs. butterfat in 180 days.
Daughter of Pride's Olga of Elm Shade Farm.

Ten Reasons Why a Farmer Should Use a Silo

1. As the principal business of farming is to make money, the silo on the stock farm will best accomplish this end.

2. Corn silage is the farmers cheapest source of digestible carbo-hydrates. As this is the principal food element required by our domestic animals it certainly is wise to provide it.

3. Summing up all the experiments conducted by some twenty-three different state experiment stations over a period of fifteen years, silage has proven the most economic food for the production of stock and stock products.

4. There are about one-half million silos now in use in the United States, and it is difficult to find a user who is not more than pleased with the results from his silo.

5. The greatest expense connected with the live stock business is the cost of feeding the animals. The silo will lower this cost and therefore it should be one of the first considerations of the stock keeper.

6. The silo stands ever ready as an insurance against many of the common crop failures, especially damage from early frost, drought and hail. Wet seasons often prevent the putting up of a crop of clover or alfalfa; this can be saved in the silo. Whatever is grown in the way of forage can be siloed and preserved for future use.

7. The silo is a labor saving equipment and it saves in storage space. Eight times more feed can be stored in the silo than in the mow. Cattle can be fed quickly and easily from the silo as it is in a condensed form, close to the point of feeding and is always in condition for feeding.

8. Silage is a succulent, grass-like feed, easily digested, and seems to stimulate digestion. It has much the same effect as grass, giving thrift to the animal; and less sickness is experienced among the stock when good silage is fed. Silage stimulates the milk flow and all milking stock should receive it. Silage is cooling and appetizing and it prevents many of the troubles resulting from the over feeding of concentrates.

9. Experiments and experience have taught that the most profitable feeding is liberal feeding, such as will supply the animal with its requirements. Quick growth is profitable growth, large production is profitable production, and the feeder of silage is more inclined to feed well, which means profitable feeding.

10. Competition is keen in all lines of industry and the stock keeper with a silo has an advantage over his neighbor without one. In order to compete with the silo keeper, all stock keepers must use silos. World competition in growing stock and producing stock products will require the American farmer to use the best and most economical means. With the silo we need fear no competition from any country of the world.—A. L. Haecker.

CARE OF THE BULL

When one has purchased a bull of the conformation and breeding desired, the main idea will be to so use and care for him that he will be a sure breeder and will reproduce in his offspring the desirable characteristics possessed by himself. To do this the bull must be kept in the pink of breeding condition and be the personification of masculinity and vigor. To keep him in this condition he must be well fed and cared for. His feed should be nourishing and of a character to produce muscle and red blood rather than fat. The best of hay and grain should be given him, and if his hay may be alfalfa or clover, so much the better. He should not be fattened as though for market yet he should be kept in good flesh at all times.

Most farmers prefer to allow the bull to run with the herd. This, in the opinion of the writer, is not the better way, for several reasons. First, under such conditions the bull does not usually get the extra care that he deserves. Second, he wastes much of his energy uselessly and becomes a half-hearted server. Third, it is impossible to keep any record of breeding dates, and this is essential where any systematic operations are sought. Fourth, the bull soon becomes breachy, and is a constant source of trouble and worry to the owner as well as his neighbors, and frequently becomes unruly and vicious. A much better way, it seems to us, is to confine him to a yard or paddock and use him as a breeder only when desired, and where he may be given good care and attention. When so cared for he will usually be quiet and gentle, and his vigor may be kept up to the highest point and he will sire many more strong, desirable calves than where he has the run of the herd. A small pasture where he may graze and exercise is also most desirable. Never, under any conditions, confine a breeding bull to a stall in the barn. He positively must have exercise, and running with the cows is far better than close confinement.

The ordinary farm bull does not receive the care that he deserves. Remember that he is half the herd and often more, and that good care will pay and pay well, in dollars and cents.

CLERMONT COUNTY JERSEY CATTLE CLUB
FRANK WALMSLEY, Secretary - BRANCH HILL, OHIO
Write Me for List of Animals for Sale in Our County.

Clermont Hill Farm
Walter Bee & Son
BETHEL - - OHIO

Breeders of

High Class Typy Producing Jerseys

———:———

Our Herd is Rich in the Blood of

Raleigh and Noble of Oaklands

We Believe This Combination Is the Best In the Breed

Correspondence Solicited

C. W. Ross & Son
FELICITY - OHIO

Breeders of

High Class Registered Jerseys

Rich in the Blood of the

Raleighs

And other popular breeding.

———:———

For Choice Young Stock
See Us

Herd Fully Accredited

CAMDEN TERRACE FARMS
C. G. Drewery, Owner.   MILFORD, OHIO

Our Herd is Rich in the Blood of MAJESTY
——— The Ultimate in Fashionable Jersey Breeding ———

We endeavor to maintain at all times only such matrons as exemplify our ideals of true Jersey type and capable of easily qualifying for the Register of Merit.

We are especially proud of our four Register of Merit daughters (two of which are Silver Medal cows) of IMPORTED OXFORD MAJESTY 2nd 152182.

Also our daughter of Gamboge's Oxford Majesty of F., and our choice females sired by the following: Imp. Oxford You'll Do, Imp. Golden Fern's Noble, Imp. King of Hamble, Rochette's Jolly Sultan, Sybil's Gipsy King, and other noted sires.

Our Herd Sire
TIDDLEDYWINK'S MAJESTY'S LAD 167989

Has 31¼% of the blood of the Gold Medal Grand Champion sire, Imported Royal Majesty, one of the most famous bulls of the breed, founder of the noted Majesty family of Jerseys and great grandsire of Sybil's Gamboge, who sold for $65,000 (the record price of the breed).

Occasionally we reserve for breeding purposes a good bull or two, and at this time offer the following bulls at reasonable prices:

Fern's Distinction Lad 183605
Born May 5, 1919.
Sire—Rochette's Jolly Sultan 91998.
Dam—Lad's Distinction Fern 400327. Unfortunately this splendid young cow died before we could test her.

An Unregistered Bull Calf
Born May 25, 1922.
Sire—Tiddledywink's Majesty's Lad (above).
Dam—Blue Belle's Mamie, Class AA, R. of M., 383.93 lbs. butter fat, at 5 yrs. 11 mos. old.

——— Personal Inspection Invited at All Times ———

CAMDEN TERRACE FARMS   -   -   Milford, Ohio

Activities of the Licking County Jersey Cattle Club

By FRED H. STEVENS, Sec'y.

The Licking County Jersey Cattle Club was organized in March, 1919. The membership numbering twenty-three first year.

The first annual picnic and cattle show was held at Fair Grounds in July, 1919. There were thirty-one head on exhibition.

The second annual picnic of the Club was held in August, 1920, at the home of W. C. Hall. This was a fine picnic, and a large crowd was present.

Two calf clubs were organized this year, one bunch of the calves coming from Belmont County.

The Club had a fine show at the County Fair in 1920. Sixty-some head were on exhibition, all from Licking County.

The first annual banquet was held in April, 1920, which was a real success. L. J. Taber was the principal speaker.

The first annual sale was held in October, 1920. The sale amounted to nearly $9,000, the highest priced animal bringing $370, and remaining in the county.

Our membership increased to fifty-four in 1920.

The second annual banquet was held in April, 1921, with R. D. Canan as principal speaker. A good crowd was present.

The annual picnic in 1921 was held at the Fair Grounds in August, with Prof. Erf, of the O. S. U. as principal speaker.

The second annual sale of the Club was held in October, 1921, amonting to nearly $6,000, $300 being the highest price.

Our membership increased to 70 during the year 1921.

Annual banquet was held in April, 1922, with the largest crowd of any banquet held. L. P. Bailey was the principal speaker.

This year, in June, the Club made a tour of the county, visiting about sixty herds belonging to the different breeders. The Club spent two days on this tour and it certainly was time well spent for everybody. The crowd numbered about 80 the second day.

Our fourth annual picnic was held at the Fair Grounds, Wednesday, August 9, 1922, an immense crowd being present.

The first Register of Merit calf club in the U. S. was organized by the Club this year, calves being secured from members of the Licking County Jersey Cattle Club.

The third annual sale of the Licking County Jersey Cattle Club will be held at the Fair Grounds, Newark, Ohio, Tuesday, October 17, 1922.

CAN I CASH A DEAD MAN'S CHECK?

"John Jones owed me $100," says one of our old subscribers, 'and gave me his check for that amount. I knew Jones' check was good for that much and a lot more—didn't present it for nearly a week. Jones had died in the meantime and the bank refused to pay. Will the law uphold the bank in this?"

Yes. The general rule is that a check is revoked by the death of the maker, and the authority of the bank to pay ceases on his death.

WHO OWNS THE FISH?

Writes R. S. P.: "A small stream or 'creek' as we call it in our part of the country, runs through my farm from north to south. X. Y. Z. owns the farm on the north, and, when the fish are biting good, he gets in a boat on his farm, floats down on mine, catches all the fish he can pull out and paddles home. Can I stop him?

Certainly. The ownership of the soil under a non-navigable river carries with it the right to fish, and the owner has an action of trespass against any party interfering with the right.

IF YOU BUY A LIFE INTEREST

The following question comes from S. P. D., Iowa: "A willed a piece of land to his wife, B, for her life, then to go to his son, C. I bought the widow's life interest some years ago, and this spring put in a big crop on the farm. The widow died day before yesterday and C stopped on his way home from the funeral and told me to move off. What can I do?"

Stay where you are and harvest the crop. You "sowed in peace" and you can "reap in peace."

CLERMONT COUNTY JERSEY CATTLE CLUB

FRANK WALMSLEY, Secretary - BRANCH HILL, OHIO

Write Me for List of Animals for Sale in Our County.

For NOBLES Consult

Rufus Cox

Breeder of High Class

REGISTERED JERSEY CATTLE

Manchester, Ohio

My Herd Sires are:

IVY'S OXFORD FERN 181572

A son of Imported Peerless Jolly Fern 143653, a prize winning son of Imp. Golden Fern's Noble, sire of 102 tested daughters and 55 producing sons. Dam, Oxford Ivy of Oaklands, is by Agatha's Oxford Noble, a prize winning son of the great Noble of Oaklands, whose 115 producing sons have 649 tested daughters, and out of Oxford Ivy, P. 14125, H. C., a tested daughter of Sultan of Oaklands 78745.

MASTERMAN'S NOBLE MONARCH

A son of the unbeaten champion winner over the Island of Jersey, a grandson of Combination of St. Saviour's and the grand champion Golden Fern's Noble. Dam, Monarch's Juillette, a splendid large milking granddaughter of Noble of Oaklands and Trial 2nd of Oaklands.

For prize-winning, producing Jerseys at reasonable prices, see me before you buy elsewhere.

Accredited Herd No. 16426

Rufus Cox - Manchester, Ohio

Come to Clermont County
For Good Jerseys

This is the birthplace of the ex-World's Champion,

Plain Mary

I have the highest producing herd in the county.

Monarch's Fluffy, 2 years 2 mos. old when started test, in 171 days has produced 335 lbs. butter fat.

Watch the fifty-pound list.

At present we have a few Cows, Heifers and Bull Calves for sale at reasonable prices.

Write for description.

Chas. J. Rosselot

NEWTONVILLE, OHIO

COCOTTE RALEIGH OF HILLWOOD 186388

Senior Herd Sire at Highland Farms

A Show Bull with Great Production Behind Him

::

Highland Farms is prepared to furnish bull calves sired by this great bull and out of R. of M. Raleigh dams. It will pay prospective purchasers to look over our herd and see the get of this great bull before purchasing elsewhere.

::

HIGHLAND FARMS - Branch Hill, Ohio

MRS. ALMA VOS and FRANK WALMSLEY, Owners.

ACHIEVEMENTS OF JERSEYS
The Economic Breed

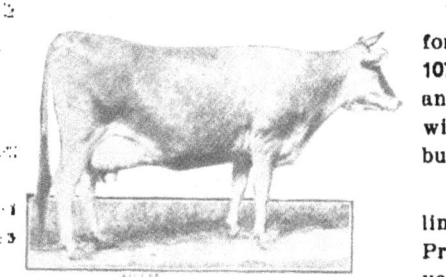
St. Mawes Lad's Lady

Vive La France

Fauvic's Prince

Financial King's Interest

Sophie 19th of Hood Farm

Grace Darling of St. Mary's

World's Champion Jersey Sire for Prepotency: **Fauvic's Prince 107961.** His first 17 daughters at an average of twenty-five months, with first calf, averaged 558.14 lbs. butter-fat.

World's Champion Jersey Yearling: **St. Mawes Lad's Lady 451568.** Produced 829 lbs. butter-fat as a yearling.

World's Champion Long-Distance Jersey Cow: **Sophie 19th of Hood Farm 189748.** (Now 17 years old.) Total production, 10 year's tests, 7037.44 lbs. butter-fat.

World's Champion Butter Cow: **Vive La France 319616.** Her first five consecutive year's records total more than the first five consecutive year's records of any cow of any other dairy breed. The total of her first five records is 4414.86 lbs. butter-fat. Her first record started at two years of age and her fifth record at six years of age. She is now on her sixth consecutive year's test.

World's Champion Jersey Cow for Twelve Years of Age or over: **Grace Darling of St. Mary's 306562.** At 12 years and 7 months of age she produced 863.90 lbs. of butter-fat.

World's Champion Jersey Cow for Production at Advanced Age and Reproduction: **Financial King's Interest 235065.** This cow is now 21 years of age and has given birth to 20 calves, 19 of which are heifers. In 24 months, starting at 18 years of age, she produced 800 lbs. of butter-fat and 3 living calves.

BREED COMPARISONS ON QUALITY OF MILK

Breed	Butter-Fat	Protein	Total Solids
Jersey	5.36%	3.80%	15.40%
Guernsey	5.00%	3.78%	14.90%
Ayrshire	3.98%	3.34%	12.75%
Holstein	3.42%	3.15%	11.80%

COME TO OHIO FOR GOOD JERSEYS —— THREE THOUSAND BREEDERS WAITING TO SERVE YOU.

THE OHIO JERSEY  NINETEEN TWENTY-TWO

Highest Jersey Butter-Fat Records

July 1st, 1922.   Class 1—Cows Under Two Years

Name and H. R. No.	Owner at Time of Test	Milk lbs.	Aver. % Fat	Fat lbs.	Age Yrs.-Mos.
St. Mawes Lad's Lady 451568	Harry D. Illif, Oregon	11,756	7.05	829.09	1—11
Lulu Alphea of Ashburn 375710	J. J. Van Kleek, Oregon	13,669	5.85	800.08	1—10
Oxford's Flower Girl 418079	G. H. M. Brewer, Oregon	11,695	5.71	667.37	1—10
Silver Chimes' Gwendola 404304	F. A. Doerfler, Oregon	10,799	5.97	644.20	1—11
Lucky Farce 298177	T. J. Foster, Penn	14,260	4.46	635.70	1—11
Owl Interest Tulip 474283	F. A. Kennedy, Vt	11,782	5.32	627.37	1—10
Lass 64th of Hood Farm 266735	C. I. Hood, Mass	9,830	6.17	606.60	1—11
Mab's Rose of S. C. 458104	Fox Bros., Oregon	10,638	5.69	605.47	1—11
The Jap's Fernleaf Queen 413813	Ainsworth Bassett, Indiana	10,084	5.93	598.04	1—10
Sophie's Harmony 431505	Mrs. Vincent Astor, N. Y.	10,290	5.77	594.04	1—11
Average		11,480	5.79	660.80	

Class 2—Cows Two and Under Two and One-Half Years

Name and H. R. No.	Owner at Time of Test	Milk lbs.	Aver. % Fat	Fat lbs.	Age Yrs.-Mos.
Pearly Exile St. Lambert 205101	Walter J. Domes, Oregon	12,346	6.61	816.10	2—5
Sophie's Bertha 313238	C. I. Hood, Mass	13,243	5.82	771.05	2—2
Sophie's Emily 352291	Wm. Ross Proctor, N. Y.	13,792	5.25	723.56	2—3
Rinda Lad's Trilby 451567	S. J. McKee, Oregon	14,314	5.32	760.79	2—5
Prince Darling's Fritzie 443318	E. L. Thompson, N. J.	13,554	5.41	732.92	2—5
Lass 66th of Hood Farm 271896	C. I. Hood, Mass	14,513	4.96	720.50	2—5
Birdie Owl of M. L. P. 415227	Robert L. Burkhart, Oregon	11,627	6.15	714.61	2—3
Benedictine Maid 421264	A. Victor Barnes, Conn	13,479	5.30	714.45	2—1
Old Man's Darling 2d 319617	Pickard Bros., Oregon	10,431	6.66	694.43	2—0
Owl's Mildred B. 414639	John R. Silby, Mass	12,410	5.58	692.03	2—3
Average		12,971	5.71	734.04	

Class 3—Cows Two and One-Half and Under Three Years

Name and H. R. No.	Owner at Time of Test	Milk lbs.	Aver. % Fat	Fat lbs.	Age Yrs.-Mos.
St. Mawes Pretty Lady 432698	L. C. Daniels, Oregon	12,550	6.57	824.33	2—10
Irene's Cherry 285828	F. D. Underwood, Wis	12,563	5.97	749.87	2—11
Waikiki's Frances 360175	J. P. Graves, Wash	11,051	6.66	735.81	2—8
Sophie's Honeysuckle 445729	Hood Farm Inc., Mass	12,038	6.04	726.88	2—8
Chieftain's Flora 389619	Woodcliffe Farm, Ohio	13,179	5.23	689.03	2—11
Koneta's Lady 2d 396741	Ayer & McKinney, New York	11,585	5.94	688.74	2—6
Princess Xenia 356699	A. V. Barnes, Conn	11,937	6.03	687.21	2—7
Lad's Lady Riotress Irene 279715	Willis Whinnery, Ohio	12,308	5.37	660.81	2—8
Lass 73d of Hood Farm 277540	C. I. Hood, Mass	19,953	6.02	659.40	2—8
Bonnie's Financial Loretta 397486	Bonnikson Bros., Cal	9,980	6.56	655.18	2—11
Average		11,814	6.04	707.73	

Class 4—Cows Three and Under Three and One-Half Years

Name and H. R. No.	Owner at Time of Test	Milk lbs.	Aver. % Fat	Fat lbs.	Age Yrs.-Mos.
Poppy's Dortha 378520	F. E. Lynn, Oregon	17,804	5.48	994.24	3—4
Vive La France 319616	Pickard Bros., Oregon	12,745	7.00	892.63	3—2
Diamond of Fair Acres 347743	John B. Stump, Oregon	17,374	4.79	831.79	3—1
Sophie's Bertha 313238	Ayredale Stock Farm, Maine	14,954	5.55	829.54	3—5
Ayredale's Over-the-Top 419983	Ayredale Stock Farm, Maine	14,029	5.73	803.45	3—3
Clara Lettie of Ashburn 368366	J. J. Van Kleek & Son, Ore.	13,748	5.80	797.12	3—3
Countess Stella of Ashburn 355367	J. M. Dickson, Oregon	11,003	7.18	777.88	3—2
Sophie's Charity 314359	C. I. Hood, Mass	11,850	6.34	751.69	3—5
Goldie's Nehalem Beauty 283330	Clifford F. Reid, Oregon	12,368	6.07	750.51	3—4
Prince Darling Edith 389918	E. L. Thompson, N. J.	14,198	5.27	747.93	3—4
Average		14,007	5.93	817.68	

COME TO OHIO FOR GOOD JERSEYS —— THREE THOUSAND BREEDERS WAITING TO SERVE YOU.

THE OHIO JERSEY  NINETEEN TWENTY-TWO

Highest Jersey Butter-Fat Records

Class 5 — Cows Three and One-Half and Under Four Years

Name and H. R. No.	Owner at Time of Test	Milk lbs.	Aver. % Fat	Fat lbs.	Age Yrs.-Mos.
Lass 66th of Hood Farm 271896	C. I. Hood, Mass.	17,794	5.11	910.60	3—9
St. Mawes' Boise Rosaire 341312	F. A. Doerfler, Oregon	14,977	5.95	891.54	3—7
Birdie Owl of M. L. P. 415227	W. M. Ladd, Oregon	14,918	5.92	883.36	3—6
Lady's Silken Glow 313311	Pickard Bros., Oregon	13,305	6.63	882.56	3—9
Eminent's Foxy Belle 304982	Jay P. Graves, Wash.	14,921	5.47	816.65	3—11
Sophie's Tormentor's Elinda 376896	Hood Farm, Mass.	14,093	5.73	807.97	3—7
Owl's Mildred B. 414639	John R. Sibley, Mass.	14,846	5.35	794.44	3—6
Sophie's Tormentor's Gem 414328	Hood Farm, Inc., Mass.	13,084	5.94	776.86	3—6
Raleigh's Corinne 346148	J. H. Humphries, Ga.	13,896	5.57	773.61	3—9
Lass 83d of Hood Farm 289023	C. I. Hood, Mass.	14,524	5.31	760.96	3—9
Average		14,636	5.70	829.86	

Class 6 — Cows Four and Under Four and One-Half Years

Name and H. R. No.	Owner at Time of Test	Milk lbs.	Aver. % Fat	Fat lbs.	Age Yrs.-Mos.
Old Man's Darling 2d 319617	Pickard Bros., Oregon	14,631	6.72	983.68	4—5
Lad's Little Pauline 349277	S. J. McKee, Oregon	15,996	5.89	941.59	4—4
Sophie's Adora 299594	C. I. Hood, Mass.	15,852	5.60	888.00	4—0
Jap Sayda's Baroness 321895	Ayer & McKinney, N. Y.	14,438	6.00	866.78	4—1
Minetta of Ashwood 330962	N. H. Smith, Oregon	16,872	5.10	860.36	4—5
Successful Queen 278743	C. I. Hood, Mass.	16,389	5.20	852.72	4—5
Princess Xenia 356699	A. V. Barnes, Conn.	13,841	6.08	840.94	4—4
Lass 64th of Hood Farm 266735	C. I. Hood, Mass.	13,445	6.08	817.71	4—5
Glen Tana Morocco Duchess 374391	Waikiki Farm, Wash.	16,125	5.04	813.10	4—4
Gwendola Olga Chimes 325849	Delmar Perkins, Oregon	15,958	5.09	812.44	4—1
Average		15,355	5.68	867.73	

Class 7 — Cows Four and One-Half and Under Five Years

Name and H. R. No.	Owner at Time of Test	Milk lbs.	Aver. % Fat	Fat lbs.	Age Yrs.-Mos.
Vive La France 319616	Pickard Bros., Oregon	14,926	6.91	1031.64	4—7
Olympia's Fern 252060	E. L. Brewer, Wash.	16,148	5.81	937.80	4—11
Goldie's Nehalem Beauty 283330	Clifford F. Reid, Ore.	15,324	5.91	904.91	4—6
Sophie's Tormentor's Elinda 376896	Hood Farm, Inc., Mass.	15,853	5.62	891.00	4—9
Fauvic Ruth 385463	A. V. Barnes, Conn.	16,919	5.19	877.70	4—6
Raleigh's Torono's Coral 2d 352298	Ayredale Stock Farm, Me.	16,825	5.21	876.80	4—11
Sophie's Bertha 313238	Ayredale Stock Farm, Me.	16,102	5.44	875.41	4—9
Sophie 19th of Hood Farm 189748	C. I. Hood, Mass.	14,373	5.95	854.80	4—11
Rosaire's Olga 4th's Pride 179509	George H. Sweet, N. Y.	14,105	5.93	836.90	4—6
Gilsand Lass 331795	David E. Moulton, Me.	11,857	7.06	836.65	4—8
Average		15,243	5.90	892.36	

Class 8 — Cows Five Years and Over

Name and H. R. No.	Owner at Time of Test	Milk lbs.	Aver. % Fat	Fat lbs.	Age Yrs.-Mos.
Lad's Iota 350672 (World's Champ.)	S. J. McKee, Oregon	18,632	5.62	1048.07	5—6
Plain Mary 268206	Ayredale Stock Farm, Maine	15,256	6.82	1040.08	8—11
Vive La France 319616	Pickard Bros., Oregon	15,272	6.81	1039.29	5—11
Lady's Silken Glow 313311	Pickard Bros., Oregon	14,939	6.95	1038.70	7—1
Fauvic's Star 313018	A. V. Barnes, Conn.	20,616	4.88	1005.90	6—11
Sophie's Agnes 296759	Ayredale Stock Farm, Maine.	16,212	6.17	1000.07	6—1
Sophie 19th of Hood Farm 189748	C. I. Hood, Mass.	17,558	5.69	999.10	7—11
Spermfield Owl's Eva 193934	Ayer & McKinney, N. Y.	16,457	6.04	993.30	8—7
Eminent's Bess 209719	W. S. Prickett, Mich.	18,783	5.13	962.80	7—2
Oxford's Wexford Spot 289464	Wm. Ross Proctor, N. Y.	16,361	5.86	958.85	7—11
Average		17,009	6.00	1008.62	

COME TO OHIO FOR GOOD JERSEYS —— THREE THOUSAND BREEDERS WAITING TO SERVE YOU.

The Long-Wished-For Publication
HISTORIES OF
FAMOUS JERSEY CATTLE

To Be Issued Serially
Edited by Harry Jenkins

Plans are completed for publishing, in a form that breeders can utilize and preserve, a series of magazines on "famous Jerseys," to be circulated as rapidly as we can prepare them, which will ultimately cover every family and include the most distinguished individuals of the entire Jersey breed.

Each number will contain 32 pages of solid Jersey facts. Ten numbers, or 320 pages, will make a convenient volume, which the subscriber may have bound and placed in his library. These 32 pages will be free from decorative embellishment and advertising, so that when bound they will form an exclusive reference book, uninterrupted by "foreign matter."

An advertising section to the magazine, in different colored paper, will be added outside the reading matter, which will serve as a "cover," to be removed when a volume is bound. (Rates for advertising space, on request.)

Sold Only by Subscription: Ten Numbers (each 32 pages), $4
320 Pages of Authentic Jersey History

Each number to be mailed separately except where a bound volume is ordered, costing $1.00 additional. (Bound volumes can not, of course, be delivered until after ten numbers are issued.) Or subscribers may return their magazines to us for binding, at the end of any volume, for $1.00 additional.

Nowhere else can the Jersey reader obtain such information—the assembled records and short histories of all the most famous Jerseys that ever lived, with a list of their distinguished sons and daughters. It represents years of work and experience concentrated upon the single purpose of providing a Jersey reference book, accurately and thoroughly.

One volume is not going to hold it all, even in condensed form. Additional volumes must follow. Read the first volume, and you will want them all. When finished, this work will amount to

AN ENCYCLOPEDIA OF GREAT JERSEYS

Subscribe Now—Do not take chances of the first numbers being exhausted. No subscriber should miss a number, and we are unable to decide definitely how many extra copies to prepare, in order to supply belated subscriptions.

The first number goes to press about Sept. 25, and will be mailed shortly thereafter. Send in your subscription before the extra copies are exhausted.

Subscription—"Famous Jersey Cattle"

HARRY JENKINS & SON, Downers Grove, Ill.: (If more than one subscription, enter amount here.)

Enter subscription for Vol. I of the magazine, "Famous Jersey Cattle," viz:

    Serial volume of ten numbers (mailed separately as issued).....$4.00 ....
                                   or
    Bound volume (mailed when ten numbers are completed)......$5.00 ....

Subscribers may have the serial or bound volume, or both, as desired. Ten serial numbers may be returned for binding, for $1.00 additional. Please strike out either item above mentioned, if not wanted. "Information for Prospective Subscribers" sent on request. Sign name in space below, with complete local address, very distinctly.

Amount of Remittance   Name...

                                  Address..

If more than one subscription is wanted, for different individuals, please write names and addresses on a separate sheet, or write a letter of instruction to the publisher.

THE OHIO JERSEY  NINETEEN TWENTY-TWO

What it Means to Have the Right Kind of Dairy Cows

The University of Illinois, after careful investigation, divided the dairy cattle of the United States into three classes, about one-third of the nation's dairy cattle in each class. The poorest third does not pay for its keep, the next third returns a very small profit, and the best third carries the other two thirds on its back, so to speak.

Average yearly production of the first one-third of these cows is 134 lbs. butter-fat. These impoverish their owners.

Average yearly production of the second one-third of these cows is 198 lbs. butter-fat. These do not pay their owners enough profit.

Average yearly production of the third one-third of these cows is 278 lbs. butter-fat. These are the strength of the dairy industry.

Average yearly production of all mature Jerseys on official test during the years 1920 and 1921 was 515 lbs. butter-fat. These are the cream of the dairy industry and 90% of these Jerseys are owned by everyday farmers.

Six of the eight world's champion Jersey cows are owned by practical dairy farmers. Of the 975 Jersey Breeders conducting official tests, over 860 are practical dairy farmers. 135 of these are in the State of Ohio.

Jerseys are the working farmers' cows.

Average yields of milk and butter-fat in a 365-day authenticated test of Jerseys received to January 1, 1922:

Age	Milk (lbs.)	Per cent.	Butterfat (lbs.)
Yearlings	6,500	5.43	353
Jr. 2-yr.-olds	6,929	5.43	377
Sr. 2-yr.-olds	7,296	5.46	398
Jr. 3-yr.-olds	7,983	5.41	432
Sr. 3-yr.-olds	8,412	5.42	456
Jr. 4-yr.-olds	8,770	5.44	477
Sr. 4-yr.-olds	9,239	5.37	496
5-yr.-olds	9,479	5.30	502
6-yr.-olds	9,902	5.21	516
7-yr.-olds	9,872	5.28	521
8-yr.-olds	9,953	5.25	523
9-yr.-olds	9,492	5.23	510
10-yr.-olds	9,754	5.22	509
11-yr.-olds	9,935	5.17	513
12 to 20 yrs.	8,902	5.31	473

Average per cent of butter-fat in 13,840 records, 5.36.

BILL NYE'S COW COPY

Bill Nye, the Humorist once had a cow he wanted to sell and he unblushingly advertised all her faults while naming the few virtues she seems to have possessed. His advertisement ran:

"Owing to my ill health, I will sell at my residence in township 19, range 18, according to the government's survey, one plush raspberry cow, age 8 years. She is of undoubted courage and gives milk frequently. To the man who does not fear death in any form, she would be a great boon. She is very much attached to her present home with a stay chain, but she will be sold to anyone who will agree to treat her right. She is one-fourth Shorthorn and three-quarters hyena. I will also throw in a double barrel shotgun, which goes with her. In May she usually goes away for a week or two and returns with a tall, red calf with wobbly legs. Her name is Rose. I would rather sell her to a non-resident, the farther away the better.

COME TO OHIO FOR GOOD JERSEYS —— THREE THOUSAND BREEDERS WAITING TO SERVE YOU.

1856   1922

The Gaumer Publishing Co.
URBANA, OHIO

PRINTERS OF FINE

JERSEY STOCK CATALOGS

Pedigree Catalog Printing a Specialty

SPECIAL PRICES TO BREEDERS ON

FINE STATIONERY

PROMPT SERVICE   GUARANTEED WORK

BOOST FOR JERSEYS

By R. D. CANAN, Fieldman A. J. C. C.

The American Jersey Cattle Club has for several years followed a constructive program for the promotion of the Jersey breed. Each year at the time of the annual meeting this program has been strengthened and improved upon. At the present time I doubt if there is a single breed association in America doing any more to promote its own particular breed than the American Jersey Cattle Club.

The efforts of the National organization has brought about some excellent results in promoting the breed but the lack of the hearty co-operation of all the breeders in this program has to some degree at least, hindered the maximum promotion of the breed.

Traveling about over the country visiting with Jersey breeders and attending the the hearty co-operation of all the breeders in meetings of the various Jersey cattle clubs, tends to give one a very definite idea as to what is really needed in order to best promote our great Jersey breed. This need in a brief way is no more than complete harmony and closer co-operation on the part of every Jersey breeder.

Jersey breeders are the ones on whom we should depend for the general boosting of the breed. However, we are not boosting when one breeder will say, 'I wouldn't take Pogis 99th of Hood Farm as a gift," another says, "If I had my way no more Jerseys would be imported to the United States," and still another is heard to say, "I wouldn't give my poorest imported Jersey for a barn full of cows like Lad's Iota." All such statements and many more of a similar nature all tend to arouse in the mind of the beginner. Is there complete harmony among the breeders of Jersey cattle? Again we have other breeders who want to register their personal opinions in the columns of the breed publication. We have read such praises as, "Ayredale Farm, Hood Farm, the Golden Glow Jersey Company and others have persistently advertised the large size, smooth type, production cow, which practically means American bred. Let us write Jersey history with the cow rather than with over advertised public auctions of foreign cattle." Such statements awaken some of our Jersey breeders to action but how are they awakened? The breeders who feel they have been improved upon, and there are many of them, immediately make reply. As a result we have several wide awake Jersey breeders expending considerable energy in carrying on a little personal discussion, regarding certain phases of the Jersey industry, which only tends to break down the morale and reduces the co-operation of all the breeders for the one great cause (promoting). The Jersey Breed.

No Jersey breeder who has studied the merits of the breed will deny the fact that the Jersey cow produces dairy products more economical than any other dairy breed. No one will deny that there are more cows with high records of production at a very advanced age in life, more early maturing cows, more cows with persistent records of high production, more cows that demonstrate the ability for transmitting breed characteristics of individuality and high production, and more cows that will adopt themselves to all extremes of climate.

The above are all accomplishments of the Jersey breed, no particular strain of Jerseys, no particular family and no particular Jersey breeder can claim a monopoly on all these wonderful achievements. The Jersey cow, the little beauty that just came into public view over two hundred years ago on the Island of Jersey and since that time has been sent to all corners of the world and developed under the conditions as she found them, she is responsible for these superior achievements.

Think back over the results and accomplishments of the Jersey breed for the past twenty-five years. Make your survey complete and you will find that the American Jersey Cattle Club, which treats all breed publicity in a neutral manner, has had considerable material of breed superiority, which might be used in promoting the Jersey breed.

In the future let us not depend entirely upon our national organization for the promotion of the breed but let every Jersey breeder forget his personal prejudices and instead work for complete harmony and closer co-operation of all the forces that can develop and promote the greatest dairy breed of them all, The Jersey.

R. D. CANAN

Watch the condition of the calf's bowels. At the first appearance of scouring or offensive odor, reduce the feed and treat for scours.

"JERSEY TYPE"
The Standard by which All Dairy Cattle are Judged

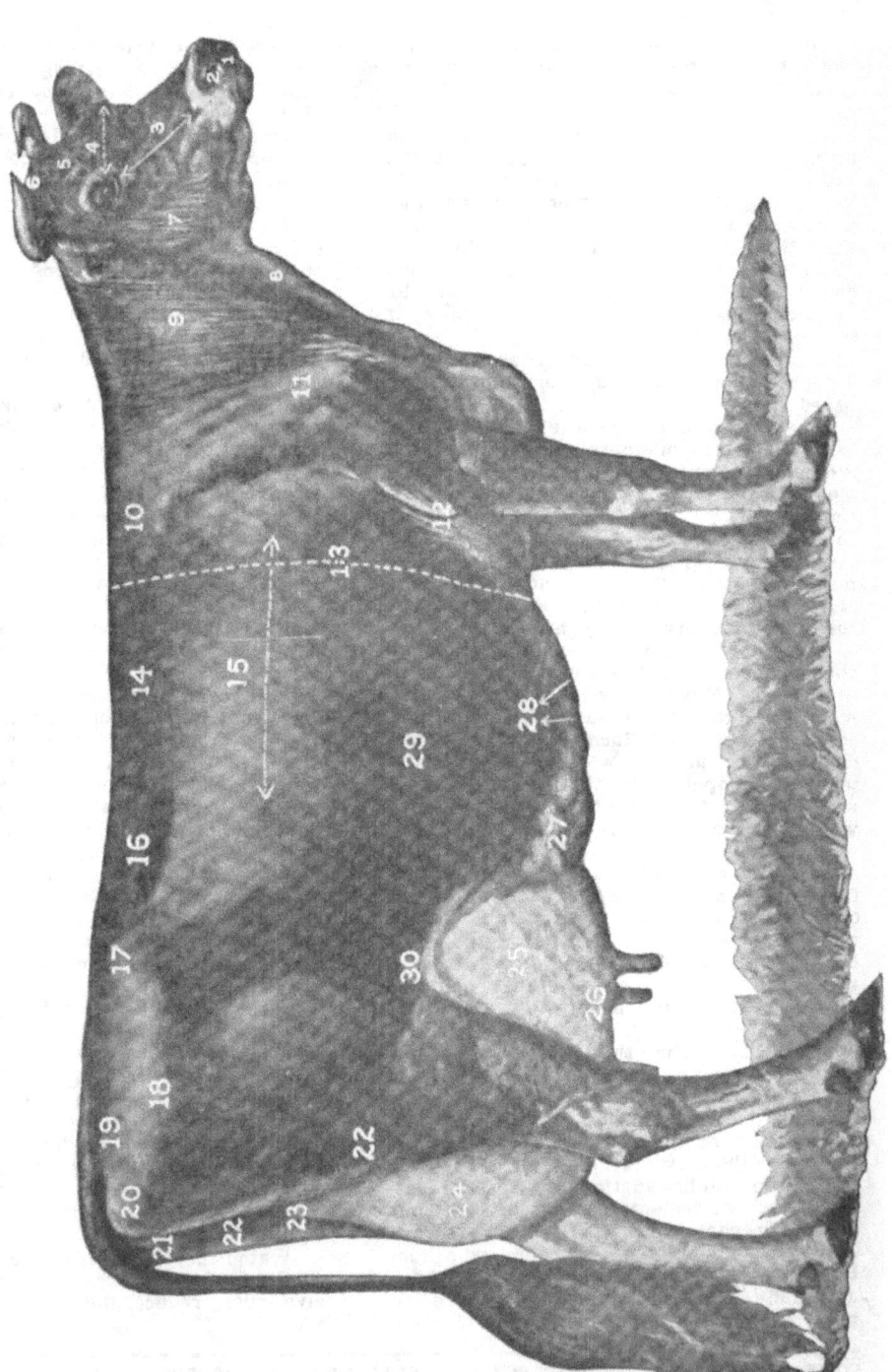

1. Mouth.
2. Nostril.
3. Length from eye to nose.
4. Breadth between eyes.
5. Forehead.
6. Poll.
7. Jaw.
8. Windpipe.
9. Neck.
10. Withers.
11. Shoulder.
12. Chest.
13. Heart-girth.
14. Back.
15. Ribs.
16. Loin.
17. Hips.
18. Thurls.
19. Tail-setting.
20. Pin bones.
21. Escutcheon.
22. Thighs.
23. Rear attachment of udder.
24. Rear udder.
25. Fore udder.
26. Width between teats.
27. Mammary vein.
28. Milk-wells.
29. Belly.
30. Flank.

THE OHIO JERSEY  NINETEEN TWENTY-TWO

Scale of Points for Jersey Cow

Adopted at Annual meeting of American
Jersey Cattle Club, May 7, 1913.

Dairy Temperament and Constitution.

Head, 7.—   Counts
  A—Medium size, lean; face dished; broad between eyes; horns medium size, incurving............. 3
  B—Eyes full and placid; ears medium size, fine, carried alert; muzzle broad, with wide open nostrils and muscular lips; jaw strong....... 4

Neck, 4.—
  Thin, rather long, with clean throat, neatly joined to head and shoulders 4

Body, 37.—
  A—Shoulders light, good distance through from point to point, but thin at withers; chest deep and full between and just back of fore legs 5
  B—Ribs amply sprung and wide apart, giving wedge-shape, with deep, large abdomen, firmly held up, with strong muscular development 10
  C—Back straight and strong, with prominent spinal process; loins broad and strong............... 5
  D—Rump long to tail-setting, and level from hip-bones to rump-bones 6
  E—Hip-bones high and wide apart.... 3
  F—Thighs flat and wide apart, giving ample room for udder........... 3
  G—Legs proportionate to size and of fine quality, well apart, with good feet, and not to weave or cross in walking 2

H—Hide loose and mellow........... 2
I—Tail thin, long, with good switch, not coarse at setting-on 1

Mammary Development

Udder, 26.—
  A—Large size, flexible and not fleshy 6
  B—Broad, level or spherical, not deeply cut between teats............ 4
  C—Fore udder full and well rounded, running well forward of front teats 10
  D—Rear udder well rounded, and well out and up behind.............. 6

Teats, 8.—
  Of good and uniform length and size, regularly and squarely placed 8

Milk-Veins, 4.—
  Large, long, tortuous and elastic, entering large and numerous orifices 4

Size, 4.—
  Mature cows, 800 to 1,000 pounds 4

General Appearance, 10.—
  A symmetrical balancing of all the parts, and a proportion of parts to each other, depending on size of animal, with the general appearance of a high-class animal, with capacity for food and productiveness at pail.................... 10
                                                                                   100

5c OR $5.00

Sec'y. Ohio Jersey Cattle Club,
Westerville, Ohio.
Dear Sir:—

Your 1921 "The Ohio Jersey" came to me free, and I doubt if $5.00 would buy my copy if I could not secure another.

The advertising in Jersey Bulletin of your 1922 Ohio Jersey for only 5c appeals to me. I'm hurrying this 5c to you.

Thank you for my copy, which I'm sure will prove both interesting and a good investment to keep as a valuable reference.

We are enjoying the A. J. C. C. Cup in 1922, and our county, Pontotoc, is trying not to lapse from 1921 efforts, though I can't say which state may have the honor of the cup in 1922.

Thank you.
                Very sincerely,
                      W. L. Thomason.

Ed. Note—The above is characteristic of the many requests for copies of the 1922 issue of "The Ohio Jersey."

Note Mr. Thomason states "We are enjoying the A. J. C. C. Accomplishment Cup in 1922." It is freely predicted that when next awarded, said cup will find its home in the state of Ohio.

COME TO OHIO FOR GOOD JERSEYS —— THREE THOUSAND BREEDERS WAITING TO SERVE YOU.

LICKING COUNTY JERSEY CATTLE CLUB

GEO. BROOKS, President,
Alexandria, Ohio.

FRED H. STEVENS, Secretary,
Newark, Ohio.

—::—

The leading families of breed represented by the breeders of this club.

Seventy-five Members

Third Annual Sale of the Club
Will Be Held on

Fair Grounds, Newark, Ohio, Tuesday, Oct. 17, 1922

Write Sales Manager for Catalogue.

Col. Weikert, Auctioneer.

Chester Folck, Sales Manager,
Springfield, Ohio.

VERBAL AGREEMENT WILL FIX A BOUNDRY

Writes A. B. B., Pennsylvania: "Six years ago my neighbor and I agreed on the line between our farms, built a line fence at our joint expense and we've both worked up to that line ever since without a word being said one way or the other. Now my neighbor says the line's three feet over on him and wants to move it back on what's always been considered my land. Can he do it? He don't deny for a second that we agreed on the line, but says that it isn't binding because it was not in writing."

Your neighbor is wrong. The general rule is that a sale of land or any interest therein must be in writing, but the courts have also held that a mutual agreement establishing a line fence is not a "transfer" of land, and, if acted upon and recognized by the parties, is binding without any written agreement.

This rule has been laid down by the Courts of your own state—also by the Arkansas, California, Connecticut, Delaware, Illinois, Indiana, Kentucky, Mississippi, Missouri, New York and Tennessee Courts.

HOW LONG TO FEED SILAGE

Cows can be fed silage for 365 days in the year without any danger of injury. The only possibility of damage is that the feeder may be careless and give them spoiled silage or the cows may not be allowed a sufficient amount of fresh air and exercise to maintain their good health. In this case the fault would be with the feeder and not with the silage.

There are many cows in the United States that receive silage all the year thru, even though the owner may have some pasturage. In many dairy districts pasture is not good for more than two months and in these cases silage is found to be an excellent supplement. Many farmers are equipped with silos for winter feeding and for summer feeding. For summer feeding it is desirable to have a silo of somewhat smaller diameter, as ordinarily the cows are not fed quite as much and in addition the silage should be fed off at a somewhat deeper daily depth than in winter in order to insure against molding from day to day.

JERSEYS ARE THE WORKING FARMERS' COWS

Seven of the eight world's champion Jersey cows are owned by practical dairy farmers.

515 lbs. of butter-fat was the average of all the records made by mature Jersey cows in 1920 and 1921. 90% of these are owned by every-day farmers.

1200 Jersey cows (all ages) on official test have averaged over 60 lbs. of butter-fat for the month of May, 1922. 70% of these are farmers' cows.

COUNTY CLUBS--WAKE UP!
Take a Hint from the Columbiana County Jersey Breeders Association

By S. N. VAN BLARICOM,
Director Ohio Jersey Cattle Club,
Salem, Ohio.

The Columbiana County Jersey Breeders Association, as it now exists, was organized some eighteen months ago and is the result of some experiments along the line of breed organization that punctuate history at irregular intervals for the past ten or twelve years.

The members of the present organization feel now that we have incorporated into our plans and specifications the elements that will build and maintain an organization worthy of the great interest it represents in the dairy industry of this country.

In the beginning it was felt that we should have some definite census figures on the number of breeders and cattle in Columbiana County, as well as all the information that could be gathered as to their quality. Accordingly a questionaire was prepared by the directors and sent out to every Jersey breeder known in the county. One hundred and ninety-seven of these were returned properly filled out, from which these interesting and unusual facts were gleaned. These 197 breeders own 2185 pure bred cattle. Of these 1566 have been tested for tuberculosis; 150 are Register of Merit cows; 10 average 682 lbs. fat; 113 average 503 lbs. fat; 114 pure bred bulls are in service, 30 of these with Register of Merit dams average 665 lbs. fat. We challenge any county in the United States of equal area to show an equal or greater number of as uniformly high qality Jersey cattle.

The next important step undertaken by the association, was to encourage the county fair exhibit. A committee was appointed to visit breeders to encourage them to place cattle at the fair and assist them in making selections. The result of this effort exceeded all expectations. While in past years the breed has been represnted by a dozen or fifteen head of indifferent quality, last year through the efforts of the association, 79 head were put in the ring, a goodly per cent of which were state fair stuff. It jarred the county fair board to the extent that they have awarded a contract for new cattle barns for this year.

Again through the efforts of the association, and our county agent the only official cow testing association in the United States was formed and started work just twelve months ago. Not much will be known of these records before 1923 as but few of the cows will have finished before that time.

The finances of the association are taken care of largely through its sales department. A graduated commission is charged ranging from 5% to 10%. This gives sufficient revenue so that those working for the association can be fairly compensated for their time and expense, also paying for the needed advertising.

The following rules and regulations of the association will interest the Jersey public. A representative of the association will meet all prospective buyers at the train when notified, conduct them over the territory and assist them in every proper way in selecting the stock wanted. Also assist in loading. All free of cost to the buyer except in case of car lots, when buyer will be expected to pay for extra fixtures in car. The association will be responsible for all transfers of cattle bought through its representative and will act as an arbitrator for any difference that may arise.

"Where do we go from here!" Straight ahead. If the success of the past is a safe criterion — limited as it is — the association is valuable to its membership. "Lets Go!"

FEEDING CALVES

Here are the nine commandments in dairy calf feeding:

Always weight the feed. Don't guess.

Avoid overfeeding. Overfeeding is sure to result in scours.

Be scrupulously clean. Clean pens, clean bedding and clean feed fed in vessels that are washed and sterilized daily are absolutely necessary.

Give the calves plenty of clean water to drink.

In feeding milk or gruel, use a thermometer; don't guess at the temperature.

Be regular in the various operations performed in caring for the calves.

Tie the calves up so they can be fed separately.

BELMONT COUNTY JERSEY CATTLE CLUB

J. R. HAINES, President,
Colerain, Ohio.

J. S. BAILEY, Secretary,
Tacoma, Ohio.

BELMONT COUNTY
Has Grown Into
One of the Leading Jersey Centers in America

———::———

HISTORY

The first Jersey Cattle came to Ohio in 1868. Purchased by the late James Edgerton of Barnesville, Belmont County, from the herd of William Crozier, Northport, R. I. Mr. Crozier was one of the first leading prominent breeders in the United States, so Belmont County can justly claim prominence as a Jersey center since 1868, with a steady growth, winning favor by the quality of her products, persistency and economical production, under the management of practical farmers.

Belmont County has grown into one of the leading Jersey centers of America, made famous by the high record made by the Barnesville and two other Cow-Test Associations. Her Champion and Register of Merit cows, Jersey Cattle "on every farm," the most progressive County Jersey Cattle Club, protecting the buyer as well as the seller, and conceiving the idea and first placing in the milk dealers' plants a milk tester hired and paid by the producers.

Our prices are moderate, within the reach of all; based on the earning values in production and offspring.

We challenge the world to offer better proof of real values—Cow Test and Register of Merit records—a fairer method of selling stock.

Many high grade cattle now for sale at bargain prices, our farmers changing from grades to pure-breds. Highly bred registered bulls, $50 up; registered females, one or a carload. Write for literature, but better come, see our cattle and methods, enjoy our hospitality.

———::———

Belmont County Jersey Cattle Club,
J. S. BAILEY, Secretary,
Tacoma, Ohio.

COUNTY JERSEY CLUBS IN OHIO

Belmont County Jersey Cattle Club—President, J. R. Haines, Colerain, Ohio; secretary, J. S. Bailey, Tacoma, Ohio.

Butler County Jersey Cattle Club—President, R. T. Shepherd, Hamilton, Ohio; secretary, W. C. McKenney, Hamilton, Ohio.

Central Ohio Jersey Breeders Association—Secretary, Earl Garlinghouse, Galena,

Columbus District Jersey Cattle Club—President, Frank Kahler, Plain City, Ohio; secretary, B. W. Keyes, Hilliards, Ohio.

Champaign County Jersey Cattle Club—Secretary, R. S. Smith, Westville, Ohio.

Columbiana County Jersey Cattle Club—President, A. U. Patton, Salem, Ohio; secretary, J. A. Binns, Salem, Ohio.

Clermont County Jersey Cattle Club—President, J. W. Dumford, Pleasant Plain, Ohio; secretary, Frank Walmsley, Milford,

Carroll County Jersey Cattle Club—President, J. H. Lewis, Carrollton, Ohio; secretary, E. E. Belkamp, Sherodsville, O.

Delaware County Jersey Cattle Club—President, J. L. Edwards, Sanbury, Ohio; secretary, H. J. Scott, Delaware, Ohio.

Darke County Jersey Cattle Club—President, C. R. Smelker, New Madison, Ohio; secretary, A. J. Warner, Greenville,

Erie County Jersey Club—President, Herbert Farrell, Sandusky, Ohio; secretary, Geo. Moisey, Sandusky, Ohio.

Fairfield County Jersey Cattle Club—President, Dr. Parkes, Lancaster, Ohio; secretary, E. D. Aneshansley, Pleasantville,

Gallia County Jersey Cattle Club President. V. J. Niday, Gallipolis, Ohio; secretary, D. D. MacLellan, Gallipolis, Ohio.

Guernsey County Jersey Cattle Club—President, Frank Casey, Cambridge, Ohio; secretary, O. C. McMunn, Lore City, Ohio.

Hancock County Jersey Cattle Club—President, Chas. A. Rudolph, Findlay, Ohio; secretary, Glen Roberts, Findlay, Ohio.

Henry County Jersey Cattle Club—President, George E. Kryder, McClure, Ohio; secretary, R. J. Perry, McClure, Ohio.

Hocking County Jersey Cattle Club—President, W. H. Pomerene, Logan, Ohio; secretary, C. W. Skinner, Logan, Ohio.

Licking County Jersey Cattle Club—President, M. C. Harter, Thornville, Ohio; secretary, F. H. Stevens, Newark, Ohio.

Mahoning County Jersey Cattle Club—President, Roy Frederick, Poland, Ohio; secretary, H. C. Reed, Canfield, Ohio.

Medina County Jersey Cattle Club—President, Geo. Abbott, Chippewa Lake; secretary, Glen L. Ganyard, Medina, R. No. 2.

Muskingum County Jersey Cattle Club—President, Fred Elliott, Chandlersville, Ohio; secretary, J. F. Martt, Zanesville, Ohio.

Rich Hill Township Jersey Cattle Club, New Concord, Ohio—President, A. L. Crawford; secretary, Roy W. Elliott.

Morgan County Jersey Cattle Club—Secretary, Wendell Tompkins, Malta, Ohio.

Miami County Jersey Cattle Club—President, Frank Tenney, Troy, Ohio; secretary, Isaac Sheets, Troy, Ohio.

Noble County Jersey Cattle Club—Secretary, John Thomas, Whigville, Ohio.

Ottawa County Jersey Cattle Club—President, H. A. Miller, Gypsum, Ohio; secretary, Eric Gregg, Genoa, Ohio.

Portage County Jersey Cattle Club—President, D. H. French, Newton Falls, Ohio; secretary, Mrs. G. W. Strickland, Ravenna, Ohio.

Preble County Jersey Cattle Club—President, Roy C. Ross, Eaton, Ohio; secretary, F. E. Cottermann, Lewisburg, Ohio.

Richland County Jersey Cattle Club—President, Ira E. Smith, Ontario, Ohio; secretary, J. W. Cline, Mansfield, Ohio.

Ross County Jersey Cattle Club—President, Hermann DuBois, Vigo, Ohio; secretary, Walter Cory, Frankfort, Ohio.

Sandy Valley Jersey Cattle Club—Secretary, R. R. Rinehart, Magnolia, Ohio.

Springfield Jersey Cattle Club—President, Albert H. Kunkle, Springfield, Ohio; secretary, Chester Folck, Yellow Springs, Ohio.

Shelby County Jersey Cattle Club—President, T. D. Allen, Sidney, Ohio; secretary, Thos. Hassly, Sidney, Ohio.

Seneca county Jersey Cattle Club—President, Blacque F. Beck, Tiffin, Ohio.

Sandusky County Jersey Cattle Club—Secretary, C. F. Nuhfer, Elmore, Ohio.

Tuscarawas County Jersey Cattle Club—President, A. H. Boltz, New Philadelphia, Ohio; secretary, Dr. T. McDermott, New Philadelphia, Ohio.

Wayne County Jersey Cattle Club—President, M. C. Ebright, Shreve, Ohio; secretary, D. W. Bookwalter, Dalton, Ohio.

Washington County Jersey Cattle Club—President, Arthur Russell, Williamstown, West Va.; secretary, J. B. Hickman, Marietta, Ohio.

Warren County Jersey Cattle Club—President, Harry E. Stokes, Waynesville, Ohio; secretary, Z. O. Worley, Morrow, Ohio.

Wyandot County Jersey Cattle Club—President, M. B. Sterner, Upper Sandusky, Ohio; secretary, M. B. Myers, Harpster, O.

Williams County Jersey Cattle Club—President, Geo. Joice, Edon, Ohio; secretary, C. C. Creek, Montpelier, Ohio.

Wood County Jersey Cattle Club—President, O. W. Hoffheims, Bowling Green, Ohio; secretary, William Dunnipace, Bowling Green, Ohio.

COME TO OHIO FOR GOOD JERSEYS —— THREE THOUSAND BREEDERS WAITING TO SERVE YOU.

Bargain in Bulls

For a quick sale I will sell the following well bred yearling bulls at $100 each, f.o.b. Columbus, Ohio:

Signal's Gamboge
Sire—TIDDLEDYWINK'S RALEIGH, soon to be a Silver Medal bull.
Dam—GAMBOGE'S SIGNAL, R. of M., 305 days' test, 335 lbs. butter fat.

Oxford's Majesty Prince
Sire—IMP. SAINT CLAIRE'S PRINCE by Rosebay's Was Wanted," a noted Island prize winner (still on the Island).
Dam—OXFORD MAJESTY'S EXCELL 3rd, now on Register of Merit test.

J. W. JONES
Supt. State School for the Deaf
COLUMBUS, OHIO

ISLAND JERSEY FARM

GEO. A. BECKER, Owner

Kelley's Island - Ohio

Breeder of High Class

Majestys

Raleighs

St. Lamberts

A few choice Heifers for sale at reasonable prices.

Interested Jersey breeders are welcome at Kelley's Island any time they come.

TUBERCULOSIS IN LIVE STOCK
(Continued from Page 36)

were slaughtered are in all parts of the country. More than that, only about 65 per cent of the cattle and swine, it is estimated, are slaughtered each year in establishments under Federal supervision, so that about 35 per cent of these classes of animals slaughtered each year in the United States do not appear in these records. It is known also that the percentage of tuberculosis is greater in the uninspected animals. In view of these points the losses shown in the following table are believed to be scarcely one-half of the total loss throughout the country.

When animals are "retained" by the Federal inspectors on account of tuberculosis it means that some evidence of the disease is discovered and the carcass is placed aside for further examination. If the disease is found to be so slight as to render the undiseased portion of the carcass fit for food, the diseased area is removed and the remainder is passed. It will be noted that such is the case in most carcasses retained, but some loss occurs for the reason that the diseased portions found unfit for food would have a considerable value if healthy.

In the animals that are retained and when the disease is not extensive enough to cause condemnation of the entire carcass, the disease is in most cases in the early stages. Had the animals been allowed to live for possibly only a short time longer, the disease would have progressed until all the carcass would have to be considered diseased. In others the lesion of disease has become surrounded by tissue that "locks it up" and prevents it from spreading to other parts of the body. Such a condition, however, is liable to change at any time during the animal's life and allow the disease to enter other parts of the body, and also to be carried out of the body and endanger healthy cattle and swine.

On the farms from which these animals came, some of the remaining cattle and swine are probably affected with tuberculosis, or will be if allowed to remain there for a sufficient length of time. Knowing this danger, State and Federal officials, when the identity of the animals can be established, trace back as many of the shipments of diseased animals as possible, and through the co-operation of the owner try to exterminate the disease from that farm.

How to proceed to make certain that cattle and swine are free from tuberculosis:

Have a competent veterinarian apply the tuberculin test. Remove all reactors promptly, and disinfect the premises immediately after the removal of the reacting cattle.

Do not feed any infected dairy products to swine or young cattle.

Retest the herd with tuberculin once a year.

CAUSE OF TUBERCULOSIS The direct and primary cause of tuberculosis is a rod-shaped germ which can be seen only with the aid of a microscope of high magnifying power. The presence of this germ in the bodies of human beings or live stock is absolutely necessary to produce the disease. The germs of tuberculosis may also be grown artificially in proper material at a temperature of about 98 degrees F.

Outside the bodies of animals the organism is not capable of reproducing itself. When exposed to the direct rays of the sun it dies quickly—a fact to be noted in the disinfection of pastures, paddocks, and barn lots. The organism may live for months, however, when it is protected by dry manure and other materials which form a crust over it and prevent its destruction by the sun's rays. It is of extreme importance, therefore, to clean and disinfect thoroughly all barns, stalls, and other inclosures which contained tuberculous animals before healthy ones are again placed in them.

While it is necessary for the germs of tuberculosis to be introduced into the body of the animal before the disease can be produced, there are many conditions or accessory causes which make animals fall victims to tuberculosis.

Animals which are fed on non-nutritious feeds, as well as those that have too little feed, become weakened constitutionally and lose the power to resist the invasion of the organisms. Stabling animals in dark, poorly-ventilated, and dirty barns helps to spread tuberculosis among the stock whenever the germs are present. Introducing a tuberculous animal is almost sure to give the disease to healthy animals in a short time. If the healthy animals drink water from the same trough or bucket the tuberculous sputum, all the animals are in serious danger of infection. Any condition that produces constant strain upon the systems of animals, such as the continued forced lactation periods of dairy cows, renders them fit subjects for the development of tuberculosis.

HOW CATTLE BECOME INFECTED WITH TUBERCULOSIS

The tuberculous cow is the greatest source of danger to healthy cattle, and inasmuch as it can not be determined just when that animal becomes a "spreader" of the germs, unless daily microscopic tests are made of the discharges from the body, and the milk is also examined microscopically, it is unsafe to keep it with healthy cattle. No cattle from outside sources should be introduced into a healthy herd until they have been tuberculin tested and found free from the disease. Unquestionably more healthy cattle acquire tuberculosis by coming into contact with affected animals than in any other way. It has been observed frequently that cattle which stand on either side of or face tuberculous animals in barns are the first to contract the disease.

The continuous water trough in barns is also accountable to a very large extent for spreading the disease. Cattle may become infected by picking over manure infected with the germs of tuberculosis. Hay, straw, or any other feed contaminated with the germs may give the disease to animals that consume such material.

Water holes and creeks into which infected milk or the washings from infected milk cans have been dumped may also be a source of the infection. The teat siphon or milking tube, in a number of instances, has been the medium by which the disease has been conveyed from one animal to another. Calves contract tuberculosis by nursing, even for a short time, cows whose udders are affected. Calves also become infected frequently by drinking milk from diseased cattle isolated from the main herd. To be safe for feed, milk from such cows should first be heated to a temperature of 145 degrees F. and held there for at least 30 minutes, but as this method requires considerable attention to assure proper heating, boiling for a few minutes is considered a better plan.

HOW SWINE MAY BECOME INFECTED

The tuberculous cow is not only a menace to other cattle, but is also the commonest source of infection to swine. In some parts of the country, especially where there are whole-milk creameries and skimming stations, feeding mixed skim milk to swine is a common practice. In that way the skim milk from one farm may be fed to hogs on another. Thus it is possible that milk from a few tuberculous cows may set up the infection among swine on many farms.

Milk is a good medium for the development of the tubercle bacilli, and swine seem to be extremely susceptible to tuberculosis. Numerous instances are on record, also, in which the whole milk is separated on the farm, the cream shipped, and the skim milk fed to swine. Consequently one tuberculous animal that is passing the germs in the milk secretions may give the disease to any or all of the animals to which any of the milk is fed. Investigations made by the Bureau of Animal Industry show that in practically every instance where tuberculosis exists among cattle, and swine are kept on the same farm, some of the latter are tuberculous. Eradication of tuberculosis from cattle, it is believed, will greatly reduce its prevalence among swine.

Another common practice of feeding, especially in the Corn Belt States is to allow hogs to run with cattle in the feed lots or pastures. If the cattle are tuberculous and the feces contain the germs of tuberculosis, in all probability the swine will contract the disease. Swine may contract tubercu-

(Continued on Page 69)

LOCUST LAWN STOCK FARM

J. R. SCHOTT, Prop.

WESTERVILLE - - OHIO

Majestys   Oxfords
Gamboges

Herd Sire:
Golden's Royal Oxford 206783

Sire—GIPSY'S ROYAL OXFORD 164809, son of Sybil's Gamboge, the $65,000 sire, out of daughter of Gipsy Majesta, two Gold Medals on Island, by a son of the great Oxford You'll Do.

Dam—GAMBOGE'S GOLDEN FANNIE 409124, Class AA R. of M. record, 11,019 lbs. milk, 750 lbs. of butter, as a 3-year-old.

At All Times
Visitors Are
Welcome

Locust Lawn Stock Farm

Woodbrook Farm Jerseys

Bred and Owned by
HORACE L. BEVAN

WILMINGTON - OHIO

For Sophie's Tormentors See Me

Under ordinary farm care and on two milkings daily my cows produced as follows:

Cow	Milk	Average % Fat	Total Fat
Torono's Kaleeta (5 yrs. 3 mo.) (360 days)	13,926	4.53	630.21
Torono's Starlight (6 yrs. 1 mo.) (359 days)	10,352	5.42	560.98
Torono's Pansy of Wood Brook (3 yrs. 8 mo.) (365 days)	8,989	5.63	503.20
May Signal's Nina (7 yrs. 4 mo.) (355 days)	11,117	4.49	499.04
Pogis Torono's Pollyanna (2 years) (361 days)	10,338	4.78	493.36
Pogis Torono's Tot (3 yrs. 3 mo.) (365 days)	9,048	5.01	453.17
Pogis Torono's Flo (3 yrs. 10 mo.) (365 days)	7,475	6.04	451.84
Pogis Torono's Rose (3 years) (343 days)	7,207	5.63	405.45
Starlight's Tulip (1 yr. 10 mo.) (365 days)	6,289	5.93	372.79

Bonny View Jerseys

A Fully Accredited and Register of Merit Herd

Herd Sire:
MUDGIE'S FAIRY LAD 156451

A great sire, combining the blood of Gamboge's Knight, Raleigh's Fairy Boy and Sensational Fern.

The daughters of this great sire are now being placed on Register of Merit test as they freshen:

FAIRY LAD'S DAISY BUTTERCUP 452166, began test as a junior 2-year-old on Nov. 4th, 1921, and has produced in nine months 7645 lbs. of milk, 434 lbs. of butter fat.

FAIRY LAD'S MILKMAID 471653, began test as a junior 2-year-old Dec. 1st, 1921, and has produced 6229 lbs. of milk, 393 lbs. of butter fat.

These records made with ordinary farmer's care, attention and feed.

My Females are rich in the blood of Golden Fern's Lad.

I have one choice Yearling Bull and two Bull Calves to sell at reasonable prices.

COME TO BONNY VIEW FOR GOOD JERSEYS

Charlie Le Galley, Owner
BOWLING GREEN, OHIO

To improve your dairy herds use OHIO-BRED JERSEY SIRES.

To get the greatest improvement in your herds and the most profitable production from them use **Sugared Schumacher Feed** as the carbohydrate part of the ration for your cattle of all ages and all stages of lactation.

The Quaker Oats Co.

TUBERCULOSIS IN LIVE STOCK
(Continued from Page 67)

losis also by eating parts of the carcasses of infected cattle, swine, or poultry. Other sources of contamination are infected sputum from human beings, and the feeding of uncooked garbage containing the germs of tuberculosis. Tuberculous swine, like diseased cattle, may also infect one another.

SYMPTOMS OF TUBERCULOSIS It must be understood that tuberculosis is a disease which often gives no indication of its presence by external symptoms. Yet persons skilled and experienced in dealing with the disease among animals frequently are able to detect certain abnormal conditions which lead them to pronounce the animal as probably affected with tuberculosis. A generally run-down condition, accompanied with a cough, is often considered to be an indication of tuberculosis, but is not a conclusive symptom. When tuberculosis is suspected it is always advisable to apply the tuberculin test without delay.

As the disease often involves the lymphatic glands in various parts of the body, an examination of such glands as can be felt in the living animal is sometimes helpful in diagnosing the disease. The glands of the throat, udder, and point of the shoulder often present an abnormal condition, such as an enlargement or hardening. Animals affected with tuberculosis in advanced stages often show a "staring" coat and a generally unthrifty condition. When the throat glands of an animal are affected, it often holds its head in an abnormal position in order to relieve the pressure which causes difficult breathing. Increased respiration is often noted when the lungs or lymphatic glands of the thoracic cavity are affected. When some of the glands of that cavity are extensively diseased, the animal often develops bloat. Diarrhoea is often evident in some cases in which infection has extended to the abdominal cavity. The symptoms mentioned, though typical, must not always be expected when animals are tuberculous; animals that are extensively diseased are often in apparently perfect physical condition.

METHODS OF DIAGNOSIS Microscopic examinations of sputum, milk, and bowel discharges of an animal are sometimes made to determine the presence of tubercle bacilli and to diagnose tuberculosis, but after many years of experience the tuberculin test is now considered to be the most practicable and satisfactory way of discovering the disease in the living animal. The inoculation of guinea pigs with emulsions made from milk or discharges from the living animals is sometimes resorted to as a means of diagnosis, but that method of examination is technical and requires special scientific training and equipment. Besides, cases of tuberculosis may be overlooked when laboratory methods are used, because tuberculous animals do not always discharge the tubercle bacilli.

THE TUBERCULIN TEST Testing animals with tuberculin is the process of introducing tuberculin into the animal and interpreting results according to well-known standards. Tuberculin is a laboratory product prepared scientifically and, when of standard potency and in the hands of skillful persons, it is a reliable agent for detecting tuberculosis in animals. It contains no living tubercle bacilli, but is a product of the growth of tubercle bacilli properly mixed with a substance on which it has grown and properly diluted and preserved. No harm can result to healthy animals from the proper application of tuberculin even if doses many times greater than the regular ones are used.

The use of tuberculin by untrained persons is to be discouraged for the reason that in many cases its effect on tuberculous animals is unobserved and not understood by those unfamiliar with its action. Tuberculin, by its immunizing property, can cause tuberculous animals to fail to respond to its application at another time; therefore it may be misused by unscrupulous persons.

THE SUBCUTANEOUS TEST (UNDER THE SKIN) The most frequently used method of testing is the subcutaneous test, which consists in injecting the proper quantity of tuberculin underneath the skin into the subcutaneous tissue. If an animal is tuberculous, the action of the tuberculin causes a fever, which is indicated by a rise in temperature. This rise, under ordinary conditions, may occur any time between the eighth and twentieth hours after the tuberculin is injected, but in some cases it is desirable to measure the temperature before the eighth hour and continue to the twenty-fourth hour or longer.

The temperatures are measured at least three times in advance of the injection, at two-hour intervals, to learn whether the animal is in proper condition to receive the test. The temperatures after injection are taken every two hours until the test is completed. The proper interpretation of the temperatures is made by the person applying the test, and a careful observance of any clinical changes is always important in determining the result. It can not be set forth too strongly that the test including

(Continued on Page 102)

WAYNE FARM - - Waynesville, Ohio

H. E. STOKES, Owner

Breeder of High Class Jerseys

RALEIGH BLOOD PREDOMINATES

This is one of the oldest established herds of Registered Jersey Cattle in the state. We were among the first to do official testing—which is the only way to determine the true value of the dairy cow. We are milk producers and we work our cows. Any cow that will not produce 8000 lbs. of milk annually is not good enough for us.

Production First, Last and Always

Our Herd Sire:

CORA'S RALEIGH'S JOLLY 154308

A brother to RALEIGH'S FENDORA, Premier Cow of Ohio. She made more butter fat in her life than any other Ohio cow of any breed. She won one Gold and two Silver Medals for high production.

As fast as they freshen we are placing the daughters of Cora's Raleigh's Jolly on test. They show great promise.

Consult Me When in the Market for Good Jerseys

H. E. STOKES - - - Waynesville, Ohio

Watch The Owl Interests

Watch The Owl Interests

MAPLEWOOD'S INTERESTED OWL 151916
His first 10 daughters to freshen were placed on Register of Merit test; they are qualifying with creditable records. One of them is a Class 3A Silver Medal winner.

MAPLEWOOD STOCK FARM

Breeders of
OWL-INTEREST JERSEYS
Federal Accredited Herd

Young Stock for Sale at Reasonable Prices

GEO. F. ABBOTT, Prop. - Chippewa Lake, Ohio

THE OHIO JERSEY  NINETEEN TWENTY-TWO

The Gold Medal Bulls of the Breed

	Name	Owner
*1.	Golden Glow's Chief 61460	Pickard Bros., Marion, Oregon
*2.	Hood Farm Torono 60326	C. I. Hood, Lowell, Mass.
3.	Imported Oxford You'll Do 111860	T. S. Cooper & Sons, Coopersburg, Pa.
*4.	Pogis 99th of Hood Farm 94502	C. I. Hood, Lowell, Mass.
5.	Irene's King Pogis 73182	Clark & Emery, Belvidere, N. Y.
6.	Poppy's St. Mawes 115434	Ed. Cary, Carlton, Oregon
7.	Royal Majesty 79313	Edwin S. George, Detroit, Mich.
8.	St. Mawes 72053	W. A. Haynes, Eagle, Idaho
*9.	Silver Chimes of S. B. 96021	Del Perkins, Carlton, Oregon
10.	Spermfield Owl 57088	R. A. Sibley, Spencer, Mass.
*11.	The Imported Jap 75265	Ayer & McKinney, Philadelphia, Pa.
*12.	Rosaire's Olga Lad 87498	Ed. Cary, Carlton, Oregon
*13.	Valentine's Ashburn Baronet 100044	J. M. Dickson & Son, Shedd, Ore.
14.	Eminent's Pilot 75364	D. C. Dilworth, Spokane, Wash.
*15.	Rinda Lad of S. B. 89518	G. G. Hewitt, Monmouth, Ore.
16.	Pogis 75th of Hood Farm 94501	Ed. Cary, Carlton, Oregon
*17.	Sophie 19th's Tormentor 113302	Ed. Lasater, Falfurrias, Texas
*18.	Holger 109744	Wm. McBride, Shedd, Oregon
*19.	Fauvic's Prince 109761	A. V. Barnes, New Canaan, Conn.

* Gold and Silver Medal Bulls.

The Silver Medal Bulls

	Name	Owner
*1.	Fauvic's Prince 107961	A. V. Barnes, New Canaan, Conn.
2.	Gamboge Whitie's Majesty 121493	J. G. Howland, Quechee, Vt.
*3.	Golden Glow's Chief 61460	Pickard Bros., Marion, Ore.
*4.	Hood Farm Torono 60326	C. I. Hood, Lowell, Mass.
5.	Hood Farm Torono 35th 99265	O. E. Stevens Henry, Rockville, Conn.
6.	Imported Financial Baron 139499	A. F. Chaffie, Hotchkins, Colo.
7.	Lou's Torono 106614	Fairview Farm, Inc., Atlanta, Ga.
*8.	Pogis 99th of Hood Farm 94502	C. I. Hood, Lowell, Mass.
*9.	Rosaire's Olga Lad 87498	Ed. Cary, Carlton, Ore.
*10.	Silver Chimes of S. B. 96021	Del Perkins, Carlton, Ore.
*11.	Sophie 19th's Tormentor 113302	Ed. C. Lasater, Falfurrias, Tex.
*12.	Valentine's Ashburn Baronet 100044	J. M. Dickson & Son, Shedd, Ore.
13.	Sayda's King of Meridale 121724	S. W. Bliss, St. Albans, Vt.
14.	Rinda Lad of S. B. 89518	G. G. Hewitt, Monmouth, Ore.
*15.	Gedney Farm Giel's Oxford 75998	J. G. Howland, Quechee, Vt.
16.	St. Mawes of Ashburn 115996	J. M. Dickson & Son, Shedd, Ore.
17.	Adelaide's Sultan 123005	Geo. Eldridge, Fruitland, Ida.
*18.	Holger 109744	Wm. M. MacBride, Shedd, Ore.
19.	Tiddledywink's Noble 106587	Mrs. F. Gale-Neal, Turner, Ore.
20.	Hood's Sophie's Tormentor 145709	U. S. Bureau of Animal Inds., Wash., D. C.
*21.	The Imported Jap 75265	Ayer & McKinney, Philadelphia, Pa.
22.	Gamboge Knight's Fox 106160	Meldrum Gray, Roswell, N. M.
23.	Spermfield Prince Interest 95697	F. A. Kennedy, Windsor, Vt.
24.	Meridale Prince Darling 135643	E. L. Thompson, Dover, N. J.

* Gold and Silver Medal bulls.   August 1, 1922

COME TO OHIO FOR GOOD JERSEYS — THREE THOUSAND BREEDERS WAITING TO SERVE YOU.

QUINBY FARMS
Established 1891

H. Anna Quinby, Owner - - 207 First Nat. Bank Bldg.
COLUMBUS, OHIO

BREEDER OF REGISTER OF MERIT JERSEYS
PLAIN MARY, TORMENTOR AND INTERESTED OWL BLOOD

We bred Daisy of Edenton 112761, dam of Plain Mary, 1040 lbs. fat, also owned Sylvia of Edenton 88521 and Eve of Edenton 73566, grandam and great grandam of Plain Mary. The undefeated show bull of his day, Dandy of Edenton 33912, grandsire of the ex-world's Jersey butter queen, for years stood at the head of this herd.

Today we have in our herd Goldie L. Bessie and Maggie of Beech Ridge, daughters of Pogis Boy St. Lambert 67337, sire of Plain Mary; also eight granddaughters of this famous bull and one granddaughter of Daisy of Edenton, dam of the ex-world's champion.

Our Present Herd Sire: MARY'S ROSE KING 15260

Son of Rose Girl's Fancy, 523 lbs. fat as a junior 2-year-old, out of Penobscot King, son of Plain Mary, stands at the head of our herd today.

MARY'S ROSE KING is the grandson of Rose Girl's Fancy, who has a record of 834 lbs. fat as aged cow, which gives our herd bull a record of 1874 lbs. of fat or an average of 937 lbs. from his grandams.

Young Stock from this "Bred for Butter" Bull for sale.

The Enacting Clause

We, the Jersey Breeders of Columbiana County, Ohio, in order to form a more perfect union, to procure a higher class of cattle, to provide for co-operative advertising, to promote the general welfare of our wives and our children, did ordain and establish the

Columbiana County Jersey Breeders Association

And this is why we did it: Because there are in our county nearly 200 breeders of pure bred Jerseys; not many rich men's herds, but just good working farms, where the owner is in charge of the cows; many herds running from 20 to 40 head. These men have been using the best bulls they could buy for years, and the result is they now have a surplus of extra good cows and heifers and they have saved a few bulls from their very best cows.

Owing to the many herds to select from, we can easily find animals enough of a class for carload shipment, thereby lowering transportation charges.

We are at present well supplied with heavy Springers and
Calf Club Heifers.

Come to SALEM, OHIO, and See Them

Columbiana County Jersey Breeders Association
JOHN A. BINNS, Secretary-Treasurer.

PLAIN MARY 268206, EX-WORLD'S CHAMPION JERSEY
15,256 Lbs. Milk, 1,040.08 Lbs. Butter-fat at 8 Years 11 Mos. Old

The Breeding Behind the Former World's Champion--"PLAIN MARY"

By H. ANNA QUINBY, Columbus, O.

When Plain Mary broke the world's record and was crowned the queen of the Jersey breed, Maine got the credit for her phenominal record.

Along with Maine, Ohio should receive recognition because Plain Mary and eleven of her ancestors were bred in Ohio, seven being bred in Clermont County alone.

The Quinby Farm, at Edenton, had Plain Mary's dam, Daisy of Edenton, and owned Sylvia of Edenton, Eve of Edenton and Dandy of Edenton, other ancestors of the world's champion cow.

When J. W. Dumford, a young school teacher, decided to farm, he purchased Daisy of Edenton, who proved to be a real foundation cow, having given birth to nineteen calves in as many years.

Today in the herds of J. W. Dumford, on Quinby Farm, can be found two splendid grandsons of Plain Mary, Plain Prince and Mary's Rose King, which stand at the head of these two Clermont County herds. H. Anna Quinby, Columbus, Ohio, who now owns the Quinby Farm, owns two sisters and nine nieces of this great cow.

Plain Mary is not only an Ohio cow but nearly as it is possible to be an American bred cow, because she traces back through Daisy of Edenton, Sylvia of Edenton, Lake's Niobean Jr. 19493, Lady Princess 16901, Lake No. 7816, Lilly 7th, No. 4711, to Lilly No. 1, seven generations to the very first cow registered by the American Jersey Cattle Club, "Lilly" being imported August, 1852, just 70 years ago. Plain Mary not only traces to Lilly seven times, but to Duchess cow No. 2 twice, three times to Pansy No. 3, seven times to Spendid bull No. 2, once to Pilot bull No. 3, and 72 times to first 46 Jersey cattle registered.

In the chain of seven links between Plain Mary and Lilly, Daisy of Edenton, Sylvia of Edenton and Lake's Niobean Jr. were bred and owned by Clermont County people.

On the sire's side Plain Mary is of the St. Lambert breeding, W. F. Rondebush and Brodwell and Cochran, of Batavia, having bred four St. Lambert ancestors of Plain Mary.

YOU'LL LAUGH TILL YOU CRY
When You Read
"The Colonel's Thunder"
Now Being Written by
COL. D. L. PERRY
Auctioneer
COLUMBUS, OHIO

TOM DEMPSEY
SALE MANAGER, JERSEY CATTLE
Westerville, Ohio

JERSEY CATTLE   Belgian Horses

Buff Leghorn Chickens   Duroc Jersey Hogs

CHESTER FOLCK

SALES MANAGER

R. F. D. 9, Springfield, Ohio

TELEPHONE
Bell Phone, Farmers' Line
Springfield Exchange

LOCATION
Five miles south of Springfield on the
Yellow Springs Pike; Stop 11 on
Springfield & Xenia Traction

THE OHIO JERSEY   NINETEEN TWENTY-TWO

Officers and Directors of the Ohio Jersey Cattle Club for the Year 1922

H. W. BONNELL, Youngstown, President.

R. S. SMITH, Westville, Vice President.

J. E. MORRIS, Westerville, Sec'y.-Treas.

Name	Address	Director From County
L. P. Bailey,	Tacoma	Belmont
A. I. Negus,	Bridgeport	Belmont
W. C. Shepherd,	Hamilton	Butler
Earl Pyle,	Clarksville	Clinton
S. N. Van Blaricom,	Salem	Columbiana
Chester Folck,	Yellow Springs	Clark
Albert H. Kunkle,	Springfield	Clark
Mrs. C. G. Drewery,	Milford	Clermont
E. J. Eckleberry,	Delaware	Delaware
Herbert Farrell,	Sandusky	Erie
T. P. White,	Hooker	Fairfield
O. C. McMunn,	Lore City	Guernsey
L. L. King,	Findlay	Hancock
George E. Kryder,	McClure	Henry
George Brooks,	Alexandria	Licking
Randall Anderson,	West Austintown	Mahoning
Grant Chidsey,	Brunswick	Medina
J. F. Martt,	Zanesville	Muskingum
H. A. Miller,	Gypsum	Ottawa
D. H. French,	Newton Falls	Portage
Robert Wylie,	Circleville	Pickaway
Clark A. Garber,	Bellville	Richland
Hermann DuBois,	Vigo	Ross
J. I. Myers,	New Dover	Union
D. W. Buchwalter,	Dalton	Wayne
J. D. Hervey,	Marietta	Washington
H. E. Stokes,	Waynesville	Warren
David Dow,	Carey	Wyandot

Every organized County Jersey Cattle Club is privileged to appoint one of its members as a Director of the Ohio Jersey Cattle Club.

Official Directory of Members of the Ohio Jersey Cattle Club

R. H. Adams, Cadiz, R. 2.
Randal H. Anderson, West Austintown.
D. B. Armstrong, St. Clairsville, R. 1.
A. G. Abbott, Wadsworth.
Geo. F. Abbott, Chippewa Lake.
Alcock & Lewis, Reno.
E. D. Aneshansley, Pleasantville.
J. B. Brockway, Chagrin Falls, R. 1.
A. Brandewine, Sidney.
Ellis B. Barkley, Oxford.
Blacque F. Beck, Tiffin, R. 9.
Geo. A. Becker, Kelly's Island.
John A. Binns, Salem.
Bates & Krider, Salem.
W. A. Boettner, Wayland.
Mrs. Chester K. Brooks, Mentor, Ohio.
George Brooks, Alexandria.
Otis Brooks, Morral.
F. G. Boner, Fredericktown.
R. H. Brown, Indianapolis, Ind.
Hugh W. Bonnell, Youngstown.
L. P. Bailey, Tacoma.
I. Robert Blackburn, Dayton.
F. A. Bick, Norwalk, R. 4.
W. E. Brown, Hubbard.
F. G. Bowman, Elkton.
Wm. F. Back, Georgetown.
G. Harold Burr, Lodie.
Forest Bigelow, Plain City.
Col. H. N. Barker, Salem.
Horace L. Bevan, Wilmington, R. 2.
Tom Brown, Spencer, Medina Co.
C. F. Bruning, Woodville.
M. P. Bartmess, Marietta, R. 4.
A. W. Brown, West Sonora.
Walter B. Bee, Bethel.
George K. Besuden, Foster.
Brooks & Barker, Salem, R. 4.
J. C. Brantingham, Winona.
A. C. Bailey, Quaker City, R. 4.
J. S. Bailey, Tacoma.
W. A. Black, Cadiz, R. 3.
O. J. Bailey, Tacoma.
Allen Bailey & Son, Barnesville.
H. H. Boltz, New Philadelphia.
W. H. Busic, Ashley.

Grant Chidsey, Brunswick.
R. D. Canan, Indianapolis, Ind., 600 Century Building.
A. C. Clemens, Upper Sandusky.
Campbell & Hibbs, Salem.
Vernon E. Crouse, Youngstown, R. 4.
Rufus Cox, Manchester.
J. E. Cromwell, Springfield.
W. H. Craig, Diamond.
C. C. Creek, Montpelier.
Frank S. Casey, Cambridge.
J. E. Clark, Anderson.
L. S. Calvert, Selma.
Tom Dempsey, Westerville.
Herman DuBois, Vigo.
Jas. S. Devol, Marietta.
C. W. Damon & Son, Brunswick.
James Dair, Dayton, R. 13.
Wm. Dunnipace, Bowling Green.
J. W. Dumford, Pleasant Plain.
J. Russell Deval, Marietta.
Mrs. L. D. Drewry, Milford.
Walter Edgerton, Hanoverton.
M. C. Ebright, Shreve, Wayne Co., R. 3.
Ray W. Elliott, New Concord.
J. B. Elson, Magnolia, Ohio.
Herbert Farrell, Sandusky.
W. F. Forbes, Yellow Springs.
D. H. French, Newton Falls.
C. C. Folck, Springfield.
R. E. Frederick, Poland, R. 2.
Arthur N. French, Lebanon.
R. M. Finley, New Concord, R. 4.
Wm. Finch, Dayton, R. 13.
E. H. Fitch, Hudson, Ohio.
A. O. Felton, Zanesville.
Harley E. Fowkes, Bellaire.
Ford Seed Co., Ravenna.
E. E. Gunberling, Kent, R. 7.
Mrs. E. E. Gunberling, Kent, R. 7.
J. Ganyard & Son, Medina.
E. S. Garlinghouse, Galena.
Wm. Grothaus, Collinsville.
W. C. Garber, Bellville.
Geo. M. Gray, Coshocton.

(Continued on Page 77)

COME TO OHIO FOR GOOD JERSEYS —— THREE THOUSAND BREEDERS WAITING TO SERVE YOU.

For High Class
RALEIGHS
SEE
Russell M. Hammitt
Breeder of
Registered Jersey Cattle
Duroc Jersey and Poland
China Swine

Specialist in Growing
SEED CORN and SEED WHEAT

R. F. D. 6,
LANCASTER - OHIO

I offer at this time an exceptionally fine line bred Raleigh sire 4 years old. Write for pedigree and description.

BRIGHT EYED GIRL 414364
Is One of My Cows

She won in 1921 a Silver Medal and a Medal of Merit by producing 9969 lbs. of milk, 622.04 lbs. of butter-fat, at 2 years 11 months old.
This record is but 24-100ths of a pound less fat than the National Champion Class AA for same age.
On next test I hope to make her a Gold Medal cow.

W. I. Griffith
GALENA - - OHIO

Majestys and Raleighs
Bred Heifers and Heifer Calves
My Specialty

Geo. F. Switzer
DEFIANCE, OHIO
R. D. 12

KENWARD FARM
W. F. KENNEDY, Prop.
BLUE ASH - - OHIO
(Hamilton County)
Breeder of
Registered Jersey Cattle
And
BARRED PLYMOUTH ROCK
POULTRY

My Herd Is Rich In the Blood of the World's Champion Producing Jerseys:
THE SOPHIE TORMENTORS

My Herd Sire:
Letty's King Torono 158784

Is a grandson of Hood Farm Torono, the greatest sire of the breed.

For Sophie Tormentors
See Me

Hedge Row Farm
A. G. ABBOTT, Owner
Wadsworth - - Ohio
"Hood Farm and Olga" Breeding

We Offer at This Time a Fine Yearling Son of
LETTY'S TORONO OF SHADYNOOK 166491
Whose dam has a record of 684 lbs. of butter in one year.
His dam, Lad's Olga Ann 379474, has a Cow Test Association record of 530 lbs. fat in 1 year.
This is a fine individual, guaranteed right in every way. Priced to sell.
Write for description and pedigree.

Koop's Dairy Farm
THEO. W. KOOP, Prop.
LIMA - - OHIO
My Herd Sire is:
Chief Raleigh 2nd
First prize at Ohio, Illinois, Missouri, Kentucky and Tennessee State Fairs; Fourth prize, National Dairy Show, 1921.
Write me about a good son of this great bull.

THE OHIO JERSEY  NINETEEN TWENTY-TWO

LIST OF MEMBERS, OHIO JERSEY CATTLE CLUB
(Continued from Page 75)

H. H. and F. R. Green, Johnstown.
Earl Giffin, R. 2.
T. O. Griffith, Diamond.
W. J. Griffith, Galena.
Chas. A. Geyer, Norwich.
Mrs. Geo. Gulth, Anna.
J. R. Haines, Coleraine.
J. D. Hervey, Marietta.
Wm. C. Hamilton, Glenford, R. 3.
O. W. Hoffheins, Bowling Green, R. 4.
Hamer and Lockwood, Lewistown, Ohio.
M. C. Harter, Thornville.
Charles S. Hatfield, Springfield, R. 4.
C. E. Heinaman & Son, Marion.
George D. Heisy, Newark.
W. R. Hannah, Elkton, R. 1.
Fred Heppley, Salem.
C. M. Hart & Son, Isleta, R. 1.
C. E. Harris & Son, Medina.
Haines & Cochran, Blanchester.
Russell M. Hammitt, Lancaster, R. 6.
B. W. Havens, Galena, R. 1.
F. X. Hoffmann, Millersville.
J. B. Hickman, Marietta.
H. E. Hirst, Barnesville, R. 4.
Hartman Stock Farm, Columbus.
H. L. Heck, Granville.
J. T. Irwin, Galena.
Mrs. Kate Ireland, Cleveland.
H. W. Ingersol, Elyria.
A. C. Jones, Yorkville.
S. W. Jones, Williamsfield.
J. W. Jones, Columbus, School for Deaf.
Austin Jones, Wilmington.
E. C. King, Dresden, R. 4.
Theo. W. Koop, Lima, Allen Co., R. 4.
R. J. Knapp, Lodie.
S. W. Kester & Sons, Elida.
H. S. Kelley, Mt. Gilead.
W. F. Kennedy, Blue Ash.
George E. Kryder, McClure, Ohio.
E. S. Kelly, Yellow Springs.
George H. Kelly, Mt. Gilead.
J. W. Kline, Mansfield, R. 1.
Albert H. Kunkle, Springfield.
Charles F. Kreitler, Warren, R. 7.
F. J. Kahler, Plain City.
George A. Korner & Son, Powhattan.
Corwin L. Knowles, Concord, Ky.
Albert Leedorn, St. Paris.
Charles LeGally, Bowling Green.
O. E. Lohnes, Osborn.
Col. Laughlin, Belle Centre.
J. H. Lewis & Sons, Carrollton.
T. J. Leyshone, Logan, Hocking County.
Ada M. Long, Urbana, R. 5.
Mrs. W. J. Lucius, Hamilton, Butler Co., R. 2.
Geo. C. Longworth, Felicity.
Bert Leas, Jr., Delaware.
A. W. Murphy, Cleveland, 1360 W. 3d St.
F. F. Michael, Bucyrus.
James H. Malcolm, Bucyrus.
Joe Morris, Westerville.
H. A. Miller, Gypsum.
J. F. Martt, Zanesville, R. 1.
Maryvale Farms, Youngstown, Ohio.
L. C. McCorkle, North Jackson.
M. P. Murnan, Worthington.
Lewis F. Martin, Newton, R. 1.
John Moser, Lisbon.
Norman Merwine, Westerville.
Magrew & Lispe, Westville.
N. T. Moore & Sons, Philo.
F. J. Minner, Dresden, R. 4.
S. E. Martin, Cambridge.
H. H. Miller & Son, Marietta, 1101 Thilep St.
C. M. McLaughlin, Tremont City, Clarke Co.
O. C. McMunn, Lore City.
R. L. McCorkle, Niles.
I. M. Moore, Columbus.
J. I. Myers, New Dover, Ohio.
Annie C. Newell, West Mentor.

Clyde R. Norway, Farmer.
W. R. Nelson, Springfield.
A. J. Nickols, Berlin Heights.
C. F. Nuhfer, Elmore.
George A. Nuhfer, Woodville.
J. S. Neer & Sons, Mechanicsburg.
J. E. Orr, Wayland.
Carl N. Osborne, Cleveland, Leader-News Building.
D. H. Olds, Springfield.
Earl Pyle, Clarksville.
H. N. Patterson, LeRoy.
Amos Pidgeon, Beloit.
W. O. Phillips, Centerburg.
W. S. Phipps, Freeport, Ohio.
J. W. Patterson & Son, Lafayette.
Myrtle E. Pyle, Clarksville.
E. Philbrook, Johnstown.
Delbert Perkins, Belmont, R. 2.
Hugh Patton, New Concord.
Clyde W. Pickering, St. Clairsville, R. 3.
Col. D. L. Perry, Columbus.
L. J. Powers & Son, Huron.
Miss H. Anna Quinby, Columbus, 207 New First National Bank Bldg
P. A. Race, Roseville.
C. W. Ross & Son, Felicity.
E. S. Reynolds, Dayton.
H. F. Richards, Salem, R. 5.
L. S. Richards, Salem, R. 5.
Homer E. Roberts, Mechanicsburg.
Roy C. Ross, Eaton.
E. C. Robinson, Copley.
Robinson Bros. & Clark, Plain City.
W. E. Rinehart, Magnolia.
Raymond Russell, Millbury.
J. C. Russell & Son, Williamstown, W. Va.
C. W. Reid, New Paris.
Rupert & Sons, New Waterford, R. 2.
John M. Rarey, Kenton.
G. C. and H. C. Reed, Canfield.
Ray S. Smith, Westville.
George W. Standish, Urbana.
H. E. Scot, St. Paris.
E. R. Sutz, Bloomville, R. 1.
C. R. Smelker, New Madison.
J. S. Sicker & Son, Coshocton.
H. E. Stokes, Waynesville.
C. M. Stemen, Johnstown.
G. W. Strickland, Wayland.
Mrs. G. W. Strickland, Wayland.
Homer Slagle & Sons, Poland, R. 2.
Dillwyn Stratton, Winona.
Walter Stratton, Winona.
W. H. Sawyer, Columbus, 1007 Huntington Bank Bldg.
Frank E. Snypp, Springfield.
Oliver Sparrow, Springfield, R. 4.
C. A. Stinson, Springfield, R. 5.
R. G. Sailors, Rogers.
Samuel Strawn, Logan, Hocking Co., R. 1.
W. R. Spann, Morristown, N. J.
George F. Switzer, Defiance, R. 12.
H. J. Schaback, Stockport.
J. H. Slaughter, Coshocton, R. 5.
Sheffield Farm, Glendale.
W. C. Shepherd, Hamilton, Ohio.
Miss Prudence Sherwin, Willoughby, Ohio.
J. C. Short & Sons, Xenia, Green Co., R. 9.
R. C. Smith, Lewisburg, R. 1.
A. Smith, Stevens & Son, Newark, R. 1.
Samuel Schlabach, Shanesville.
Charles Stockslager, Lewisburg.
R. C. Schwenck, Newtown.
Paul Stein, Newton Falls, R. 2.
Seth P. Scott, Lisbon, R. 5.
James T. Shaw, Lisbon, R. 4.
B. W. Stratton, Hanoverton.
Shipman & Wharton, Alledonia.
Merle D. Stacey, Marietta.
W. Kelsey Schoepf, care Cincinnati Traction Co., Cincinnati, Ohio.
J. R. Schott, Westerville.

(Continued on Page 78)

COME TO OHIO FOR GOOD JERSEYS —— THREE THOUSAND BREEDERS WAITING TO SERVE YOU.

(Continued from Page 77)

L. J. Taber, Barnesville.
P. R. Thomas, Windham.
T. B. Thomas, Wayland, Ohio.
Fred Utz, Carrothers.
S. N. Van Blaricom, Salem, R. 6.
Alma M. Vos, Loveland.
A. L. Vaughan, Mt. Gilead.
J. E. White, Greenfield.
Robert Wylie, Zanesville.
Clifford Wax, Bloomville, R. 1.
T. P. White, Hooker.
Willis Whinnery, Salem.
R. F. Whitehouse, Dayton, R. 13.
Jessie Windross, Clarksville.
Frank Walmsley, Milford, Claremont Co., R. 2.
L. A. White, Clyde.
Ervin L. Wilkinson, Brunswick, R. 3.
Charles Wittmer, Loveland.
R. G. Wolfe, Coshocton.
Brent Woodmansee, Highland, Ohio.
Z. O. Worley, Morrow, Ohio.
J. P. Walker, Gambier.
Glen Weikert, Christiansburg.
Ward Williams, Johnstown.
Allen Frederick, Poland, Ohio.
R. L. Fleming, Canfield, Ohio.
D. W. Buckwalter, Dalton, Ohio.
A. L. Sterner & Son, Upper Sandusky.
Mrs. Florence Heberding, Youngstown.
Omer C. Ryan, Greenville.
Col. B. A. McKitrick, West Mansfield.
Clark A. Garber, Bellville, Ohio.
E. F. Gates & Son, Tyrell, Ohio.
Mrs. Anna M. Mason, Loveland, Ohio.
John A. Lee, Westerville, Ohio.
Clair Metzger, Westerville, Ohio.
V. E. Crouse, Youngstown, Ohio, R. 4.
Miss Mary Dixon, Belaire, Ohio, R. 2.
L. L. Mercer, St. Clairsville, Ohio.
A. I. Negus, Bridgeport, Ohio.

— FOR —
FINANCIAL KINGS
COME TO OHIO HEADQUARTERS
GRANDVIEW FARM
E. F. and M. E. Pyle, Owners.
Clarksville - Ohio

In our select herd of Financial Kings we have a Gold Medal daughter of Financial King's Eminent 94592 and a Silver Medal daughter of San's Aloi's Niece's King 146291.

Federal Accredited Herd

We offer at this time an extra well bred young Bull. Write for prices and description.

THIS PICTURE Illustrates One of the Bill Boards Erected by the

Belmont Jersey Cattle Club

Along the National Highway

Feeding and Developing the Dairy Cow
(Continued from Page 19)

period, it sometimes becomes difficult to dry the cows. It is quite essential for the normal development of the calf that the cow should be at rest for a period of at least six weeks and preferably two months. Commence drying the cow three months previous to parturition by reducing the feed gradually, especially reducing the amount of salts given. Feed her coarse roughages such as blue grass and timothy hay. Limit the amount of water somewhat and milk her but once a day, but milk her thoroughly.

Finally reduce the number of milkings so that she is milked only on alternate days.

Care must be taken that no garget develops in the udder.

Occasionally a cow is such a persistent milker that continued attempts to dry her will injure the udder. In this case check the flow of milk as much as possible and continue milking her.

After the cow is dry gradually return to the high-grade ration with an additional amount of bone meal. When a cow has no rest a weaker calf is usually the result, but if the above suggestions in feeding are followed this is not so likely to be the case. The flow of milk after parturition will not be quite as great when the cow has not been dry, but if properly fed and cared for the cow will soon produce the normal amount of milk. Should it be necessary for any reason to force a cow dry, this can be done most efficiently by inflating the udder with pure air. This, however, is somewhat hazardous as there is danger of infection from the milk tubes.

FEEDING DURING THE DRY PERIOD 6. After the cow is dry, feeding of a ration similar to the high-grade ration should be resumed. For a short time test it is desirable to feed it in very large quantities, so that the cow will be as fat as possible at the time of freshening. However, it is better in this event to begin drying the cow five months previous to parturition instead of three. For a yearly test a cow need be only in good normal condition, that is, she could carry a sufficient amount of flesh to cover her ribs.

About two weeks previous to parturition, begin feeding an increased amount of bulky material, such as sugar beets and bran, in addition to the high-grade ration. Then gradually reduce the quantity of the high-grade ration until about a week before parturition when the ration should consist almost exclusively of beet pulp, bran and hay (blue grass or other low protein hay). For about ten days previous to freshening, two teaspoonsful of bone calcium phosphate should be added to the beet pulp and the cow should also receive one pound of Glaubers salts each day.

It is extremely necessary that a cow be in a laxative condition at this time. It might be necessary to give a dose of castor oil, about a pint, which should be given with a syringe.

PARTURITION 7. Many of the high producing cows develop milk fever. Sometimes there are indications of this even before parturition. At this time it is necessary to give the cow the greatest care and at the first signs, no time should be wasted before properly inflating the udder with oxygen or air. The severity of the case may be lessened by lightening up the feed and giving a laxative. If the udder is extremely distended before parturition, care should be taken not to relieve it too much, as this may cause a severe case of milk fever. If the udder becomes too tense, it is advisable to use a mixture of one pint alcohol, one pint castor oil and one-eighth pound of vaseline. Apply by rubbing it over the udder and massaging it gently for some time. If the teats become chapped, a mixture of arnica and cedar oil, half and half, should be applied.

FEEDING AFTER PARTURITION 8. Immediately after parturition, a little blue grass hay, some tepid water and salt may be given. On the second day, beet pulp and bran mash should be continued and oat meal added to the ration. Should be about one-half beet pulp, one-fourth oat meal and one-fourth bran.

A cow on short time test should be syringe drenched with a pint of linseed oil. Sometimes the linseed oil is nauseating and does not have the desired effect. The peristaltic action of the intestines may then be increased by using a mixture of one-third linseed oil, one-third peanut oil, one-sixth cottonseed oil, and one-sixth sesame oil with a little salt.

Do not milk the cow dry. Simply relieve her at first, if what the calf takes is not sufficient. Gradually remove the milk until about the third or fourth day when the cow may be milked dry. At this time the swelling should have subsided. Massage the udder several times a day by rubbing it with a downward motion as this will help to relieve the swelling. As soon as the swelling has subsided, begin feeding the high test ration, gradually increasing the quantity, and reducing the amount of beet pulp and bran mash.

CHARACTER OF FEED IN HIGH TEST RATION 9. For the sake of convenience, feeds are grouped according to the effects they produce. The two general

(Continued on Page 81)

MOUND VIEW FARM

Haines & Cochran, Owners - Blanchester, Ohio

BREEDERS OF HIGH CLASS JERSEYS

Rich in the Blood of

Raleigh   Oxford   Majesty

We offer no apologies for our Herd Sires:

OXFORD MAJESTY'S PATRIOT 149716

Sire—SYBIL'S GAMBOGE, P. S. 5260, H. C.
Dam—OXFORD MAJESTY'S SKYLARK 361803, a Register of Merit daughter of Imp. Oxford Majesty.

GAMBOLINE'S MAJESTY 168144

Sire—LUCINDA'S MAJESTY 107964.
Dam—OXFORD GAMBOLINE, by Imp. Oxford Majesty.

We have at all times a few select sons of the above sires, out of Register of Merit dams, to sell at reasonable prices.

Federal Accredited Herd

JERSEYS

The Economic Producers of Butter-Fat

Result of the ONLY TEST for economic production ever conducted by the U. S. Government, St. Louis, Mo.:

NET PROFIT ON   Jerseys   137%
BUTTER   Holsteins 85%
PRODUCTION   Brown Swiss 70%

The first 14 cows were Jerseys.

Result of OHIO UNIVERSITY TEST:
For each 1,000   Jersey 65 lbs. of butter
lbs. of feed   Holstein 47 lbs. of butter

Result of COLUMBIAN EXPOSITION TEST:
Net Profit Per Cow   Jerseys $52.95
for 90 Days   Guernseys 39.91
Butter Production   Shorthorns 37.92

The average yearly production of all mature Jerseys on 365-day official tests in 1920 and 1921 was 515 lbs. of butter fat.

All Records and Every Test Prove the Jersey to be the Economic Butter-Fat Cow

AMERICAN JERSEY CATTLE CLUB
324 W. 23d Street   NEW YORK

J. F. Martt

ZANESVILLE - OHIO

Breeder of

High Class Jerseys

The kind it pays the farmer-dairymen to keep.

The kind that are profitable with ordinary feed and farmers' care.

Write Me Your Wants

BONNIE DEL JERSEYS

Are easily making R. of M. under ordinary farm conditions with handsome profits on the ledger. They are Headed by

Lucky Farce's Raleigh

A splendid son of the Ohio Champion.

Buy one of his sons for production.

Norman Merwine
Westerville, Ohio

Feeding and Developing the Dairy Cow
(Continued from Page 79)

classes of feeds are known as roughages and concentrates. Roughages differ from concentrates in that they are more bulky. However, a feed may be bulky and contain a large amount of cellulose, which may have all degrees of digestibility, so that there is a "twilight zone" between roughages that are highly concentrated and concentrates that are rather bulky.

Under bulky concentrates may be classed such feeds as beet pulp, bran, cocoa bean shell and alfalfa meal. These feeds often contain the carbohydrates in such a form that a wonderful influence is exerted upon the digestion. When beet pulp is soaked it is possible to combine with it highly concentrated feeds, very finely powdered. The concentrated feeds are thus rendered more easily digested on account of the fact that the particles of beet pulp are surrounded by the meal, thus making it more easy for the gastric juices to convert the grain into chyle which is assimilated by the lacteals and conveyed into the blood. This is one of the tricks of good feeding.

Bran and alfalfa meal have an effect similar to beet pulp but are more concentrated and their carbohydrates are less digestible.

Starch and sugar feeds consist of the common grains raised on the farm and supply a large part of the digestible nutrients. These feeds contain the greatest part of the starch and sugar necessary for a good ration.

An oil and protein ration is necessary in order that the oils may be more easily supplied than when the carbohydrates must be transformed into oil.

The mineral ration is necessary to assist in digesting and assimilating the proteins.

The deficiencies are those materials which give zest and vitality to the ration and possess a stimulating effect to a certain degree.

THE HIGH TEST RATION 10. Following is the high test ration:

Bulk Ration—	Percentage composition
Dried Beet Pulp	17½
Starch and Sugar Ration—	
Wheat Hearts with Bran	
Oat meal	
Hulled Corn	47
Malted Barley	
Cane Molasses	
Oil and Protein Ration—	
Linseed meal	
Cottonseed meal	
Ground Flax	
Ground Peanut kernels	27½
Ground Soy Beans	
Ground Sunflower seed	
Mineral Ration—	
Bone Calcium Phosphate	
Soluble Blood flour	
Salt	
Calcium Carbonate	4
Iron Sulphate	
Carlsbad salts	
Deficiencies—	
Ground Millet seed	
Fenugroek	
Mustard	
Juniper berries	4
Fennel	
Anise	
Carraway seed	

Note—Thirty-one pounds of feed per day is a good average daily ration for high producers.

The basic part of the high test ration has been used for more than fifteen years in developing Ohio dairy cows, this state having by far the greatest number of high record cows of any state in the Union. While no individual ration can be entirely responsible for these records, the high test ration has been of great assistance to the feeder. In fact, it is quite necessary to have a ration that will supply all the constituents that a cow under high working conditions needs.

As will be noticed, these rations are comparatively low in protein and high in digestible fiber and mineral matter.

Experience has shown that where high protein feeds have been fed continuously there has developed a predisposition or rather a tendency toward garget and the cow has also become somewhat more susceptible to the infection of abortion. Cows that have not been fed such highly concentrated and proteinous rations at all times have proven better producers in the long run. It is true that for the short time test high protein rations are necessary but the real merit of a cow is not brought out by a short time test nor even by a yearly test, but rather by the amount of milk and butter fat that she will produce in a life time. The value of a test increases in proportion to the period of time covered.

It is sometimes of advantage to vary the high test ration and this may be done by adding a little linseed meal to the ration for a week or two, then by eliminating this and for the next two weeks introducing corn meal and in many cases the addition of beet pulp will increase the flow of milk. Slight modifications of the high test ration can be made to good advantage if the feeder is careful in studying the habits of the cow.

HOW THE HIGH TEST RATION MAY BE FED 11. All ingredients in this ration with the exception of the beet pulp are much improved by being finely ground. In fact regrinding after mixing makes a wonderful gain in the feeding value, as it rubs the oils uniformly

(Continued on Page 83)

LUCKY FARCE 298177—Ohio's Highest Tested Jersey
Register of Merit records:
14,260 lbs. milk, 747 lbs. butter, at 1 year 11 mos. old.
14,184 lbs. milk, 833 lbs. butter, at 3 years, 2 mos. old.
18,014 lbs. milk, 1105 lbs. butter, at 8 years 3 mos. old.

Lucky Farce

Is one of my good cows.

I have others and they are for sale at reasonable prices.

CLAIR METZGER, Owner - Westerville, Ohio

Huckleberries for Production

—::—

YES
AND
WITH

TYPE AND QUALITY

—::—

Federal Accredited Herd

W. H. SAWYER

Huntington Bank Bldg. - - Columbus, Ohio

Feeding and Developing the Dairy Cow
(Continued from Page 81)

into all the ingredients, reduces the fiber and makes the entire feed more digestible. The grinding of beet pulp is of no advantage as the absorption of moisture is about the same. If the pulp is combined in the ration it is desirable that the entire feed be moistened before feeding, but no more water should be added than the proportion of beet pulp in the ration will take up by itself. For instance; (dried beet pulp will readily take up and hold in suspension, two and one-half times its weight of water, so that if ten pounds of feed contains two pounds of dried pulp, not more than five pounds of water should be added to the portion of ten pounds of feed so that it does not become sticky or sloppy, but remains crumbly and permits of easy and thorough mastication). This is important as the main constituents of the high test ration are high in carbohydrates and oils and will form into a heavy compact mass if too wet. This hinders mastication, reduces the palatability and does not make for easy digestion. The practice of making the ration into a slop or semi-liquid state is not recommended as there can be no mastication and the ration is so much diluted that it reduces the capacity of consumption.

Where beet pulp is not already included in the ration, the best results are obtained by sprinkling the feed in its dry state over pulp that has been soaked in water or with coarsely ground mangles or red table beets.

The beets should be thoroughly washed and the decayed places removed, and when cut, the particles should not be finer than one-fourth inch in diameter. Sprinkling the finely ground feed flour over the soaked beet pulp or ground beets makes the feed extremely palatable and helps in the mastication of the feed and mixing with the gastric juices.

In connection with the high test ration there should be fed finely cut, leafy alfalfa hay and not to exceed eighteen pounds of well cured and highly grained corn silage. While silage is quite essential as a part of the ration on account of its diluteness in nutrients it should not be fed in large quantities. Some cows may even do better with not more than twelve pounds.

BEETS OR BEET PULP Beet pulp or beets are especially desirable in a ration as they assist in mastication and help in conveying waste materials through the intestinal tract. They aid in checking gastric fermentation and in keeping the fecal matter in a plastic condition, thus giving the intestines greater power of absorption and assimilation. While the proteins and fats are comparitively low and the fiber content high, the fiber in beet pulp is highly digestible.

When beets are used, the small red beet is the best on account of the high per cent of mineral matter which it contains. The large pink mangle probably comes next, while sugar beets and rutabagas prove beneficial but usually are of less value.

Carrots are quite desirable and may sometimes be used to replace the beets to a certain extent. Potatoes may be substituted for either beets or carrots and in sections where their cost is low it might be preferable to use them. Whenever beets or other roots are fed they should be washed and the decayed parts removed.

WHEAT HEARTS AND BRAN Wheat hearts consist of the germ of the wheat which is very high in digestible proteins, fat and mineral constituents. Wheat hearts are recovered in the first rolling process in milling operations and mixed with them is also a heavy oily layer of the shell part of the wheat berry, some of which usually goes into the middlings. It differs from bran in that bran contains more crude fiber, that is, bran retains a greater percentage of the shell and contains a smaller amount of the berry. Bran is not as good a feed as wheat hearts and the fiber is not as digestible as that of beet pulp. Wheat hearts are also a better feed than middlings or flour.

OAT MEAL Oats vary in quality to a great extent from year to year. Good oats should weigh thirty-three lbs. to the measured bushel. It is desirable to have a large portion of the hulls removed in order to reduce the amount of crude fiber which is not very digestible in the hulls of the oats. Oats with considerable crude fiber may be ground extremely fine thus making the fiber more digestible. This is likewise true of the crude fiber of other feeds. Clipped oats are much to be preferred to whole oats but best of all are oats with the hulls removed and the remainder ground into meal. Oat meal is very digestible.

CORN Corn used in the high test ration should be hulled before it is ground into meal, as the hard, cellulose part on the outside shell of the corn is extremely indigestible and is not at all essential in the ration. Great care should be taken not to feed moldy corn.

MALTED BARLEY Malted barley contains a diastase ferment which converts the starch into a form of sugar called maltose. This is extremely beneficial

(Continued on Page 90)

Feeding and Managing the Dairy Calf
(Continued from Page 23)

letting the calf have all the milk it wants. This would probably be all right if the calf were fed about every two hours, as is the case when it runs with the cow. When a young calf which has been without feed for 12 hours or more is given all the milk it will take, there is danger that it will gorge itself, thus causing digestive troubles. It is much safer to keep the quantity of feed well below the capacity of the calf than to risk over-feeding.

Care should be taken to see that any milk fed the young calves is of uniform temperature of about 90 degrees Fahrenheit. Many feeders attempt to overcome poor quality in the feed by increasing the quantity; that is, they feed more skim-milk than they would whole milk, the idea being that the added amount of the former makes up for the butter fat which has been removed from the latter. This is radically wrong. The same rules hold good in over-feeding with skimmilk as with whole milk. When, on account of age, souring, dirt, etc., the quality of the milk is poor, the quantity fed should be reduced rather than increased, because the danger from infection by such milk is very much greater than from fresh milk, and the prevention of digestive troubles should always be uppermost in the mind of the feeder. The loss of development in the calf on account of a temperary reduction in the ration when for any reason it is sour or nearly so is much less than when the digestive system becomes disordered. A calf can often take a relatively small quantity of bad milk for long periods and still hold its own or even make small gains, whereas a larger portion would endanger its life or give it digestive troubles of a serious nature.

FREQUENCY AND REGULARITY OF FEEDING Under natural conditions the young calf receives nourishment every two or three hours. In hand feeding it is best to follow these conditions as closely as possible, but because of the trouble and expense involved it has been found impracticable to feed calves more frequently than three times a day, and in some cases twice a day.

It is the practice of many dairymen to feed young calves three times rather than twice a day, because the better results obtained more than pay for the additional work. When this is done the periods between feeding should be as nearly equal as possible. The chief advantages of feeding in this manner are that the calf can not overload its stomach, and that the digestion of the feed is more evenly distributed throughout the 24 hours. When calves are fed only twice a day the utmost care should be observed to see that the feedings are, as nearly as possible, 12 hours apart. The importance of regularity in feeding can not be over-emphasized.

FEEDING DURING THE FIRST FIVE WEEKS At least four-fifths of all dairy calves are raised on separated milk, grain being used to supply the fat removed. Usually it pays well to feed whole milk for about two weeks, at the end of which time separated milk may be used in part. The proportion of the latter may be gradually increased until at the end of the fourth week it is used altogether. No fixed rules of feeding, based upon age, can be given, because the size and vigor of the calf must always be considered. Calves especially strong at birth may be put on separated milk entirely at two weeks of age, but this should not be attempted with weak ones. Until the calf is in vigorous and thrifty condition no attempt should be made to change to separated milk. This change should always be made gradually. The schedule given below for feeding calves is suggested as a guide, but it will often have to be modified to suit conditions. The supplementary feeding of roughage and grain to young calves is mentioned later under the respective headings.

For the first four days, from 8 to 12 pounds of milk from the dam should be fed. After this time the milk may be from any cow or cows in the herd, but preferably not from any that are nearly dry. Milk containing not more than 4 per cent of butter fat is considered the best for this purpose.

At the beginning of the third week the substitution of either skim or separated milk may commence at the rate of one pound a day. The quantity of the daily ration may be increased two to four pounds, depending upon the vigor of the calf. The quantity, however, should be kept well below the capacity of the calf; that is, when it does not drink eagerly what is offered, the quantity should be cut down. In most cases, at the end of the third week the ration should be approximately ont-half whole and one-half separated milk. Any increase should be made slowly so as to accustom the calf to the additional amount.

At the beginning of the fourth week, from one-half to three-quarters of the milk ration should be separated milk. During the week the change should be continued until by the end of the week only separated milk is fed. With especially vigorous calves the change to separated milk may be made about a week earlier.

After this time separated milk may be fed entirely unless the calf is very delicate. The quality fed can be gradually increased

(Continued on Page 85)

until 18 to 20 pounds a day are given. It is usually not economical to feed more than this unless milk is very plentiful.

The time that milk should be discontinued depends upon its cost in relation to the value of the calf, its breed, size, vigor, etc. The season in which it reaches the age of six months and the other feeds available at that time must also be taken into consideration. Six months is probably a good age at which to wean calves from milk. When the best of hay, silage and a good variety of grains are available, the calf may be weaned earlier than when such feeds are lacking. The season of good, succulent pasturage presents the best possible condition for weaning a calf, and when this exists the calf can be weaned earlier than when it is lacking. The stronger and more vigorous the calf the earlier it may be weaned with safety. On the other hand, the more valuable the calf the more expense the owner is warranted in incurring to develop it and the later it will probably be weaned. If skim or separated milk is plentiful, calves may be fed profitably until they are eight or ten months old.

When the calf is two months old, and if it is carefully watched, sour milk, whether whole, skim, or buttermilk, may be fed without harmful results, provided the change from sweet milk is made gradually. It is not well, howevtr, to alternate between sweet and sour.

ROUGHAGE FOR CALVES Usually a vigorous calf begins during the second week to pick at the bedding or other material within its reach, and at that time both grain and roughage of the best possible quality should be provided. Clover hay, alfalfa hay, or, if these are not available, the most palatable roughage on hand should be given the calf after the second wetk. If alfalfa is used, care should be taken that it does not cause scours; this feed should be fed sparingly at first and increased only after the calf gets accustomed to it. The essential points are that the roughage be of good quality and kept clean. The usual mistake in providing roughage is that it is not kept clean. It is either placed on the floor, where it soon becomes contaminated with the droppings, or is put into a manger or feed box, so placed that the calf can easily spoil it. Hay should be furnished at first only a handful at a time and placed so that it can not get soiled. A latticework rack, of metal or wood, is useful if it is placed high enough from the floor so that the calf can not soil the hay in any manner but still have it within easy reach. This rack can be so constructed that it can be hung on the wall or framework of the stall or pen and removed when not needed. For the first six months, at least, the calf should receive all the roughage of good quality that it will eat up clean. The quantity taken up to the time it is one month old is very small. The rack should be emptied every day and fresh roughage supplied.

Silage may be given after the calf is one month old, but the utmost care should be observed to be sure that it is fresh from the silo. For this reason very young calves, except in the hands of a very careful feeder, should not be fed silage, as it ferments rapidly when exposed to the air. Care should be used in starting with silage, using only a little at first and gradually increasing the quantity as the animal becomes used to it. Generally there is no danger that the calf will get too much roughage that is clean and of the proper quality.

If the calf has access to good pasture during the first six months it need not receive other roughage. Pastures used in summer should contain plenty of shade. It is not advisable, however, to have a calf under two months of age on pasture in the early spring.

Whole milk is "nature's balanced ration" for the calf. When skim or separated milk is fed, other feeds are used to supply the fat which has been removed. Proprietary calf feeds or meals, for feeding with separated milk, have been put on the market. Many of these have merit, but by using the feeds usually found on the farm or in markets near by the farmer can generally mix a grain ration which is as good and costs less.

Wheat bran is a grain which is readily taken by young calves. Inasmuch as one of the essential points is to induce the calf to eat grain as early as possible, bran in many cases is one of the ingredients in the mixture. Corn, a feed very commonly found on the farm, has an excellent physiological effect upon cattle of all ages and to a great extent may take the place of the fat removed from the milk. It is therefore one of the very best grains to use with skimmilk. Experiments in Virginia tend to show that corn fed to calves should be cracked rather than finely ground. Cracked corn and wheat bran therefore should be the basis of the feed mixture. Ground oats are very good for the purpose, but they are not grown on the farm so commonly as corn and in many cases cost much more per unit of feed than corn and bran. The following mixtures are recommended:

(1) Three parts cracked corn and one part wheat bran.

(2) Three parts cracked corn, one part wheat bran and one part ground oats.

(3) Three parts cracked corn, one part wheat bran, one part ground oats, and one part linseed meal.

(4) Five parts cracked corn, one part

wheat bran, one part ground oats and one part blood meal.

(5) Oats, ground.

When the calf is in its second week it should begin to receive grain, and when one month old it should eat about half a pound a day. After this time the quantity of grain may be gradually increased, feeding the calf all that it will take until three pounds a day is reached; this will probably be some time during the third month. Grain when fed with separated milk should never be put into the milk. It is questionable whether the preparation of grain in any way, such as soaking or boiling, is advisable under any circumstances. Grain should be furnished in separate feed boxes placed so that it can not be soiled by the droppings of the calf, but at the same time where the calf can get at it readily. There should be no corners in which wet feeds may ferment, and the utmost care should be taken to keep the grain fresh and clean at all times.

In dairy herds in which the entire output is sold as whole milk at high prices, there is a strong demand for feeds to take the place of the milk fed to the calves. While it is probably not practicable to take calves two days old from the cows and raise them entirely without milk, some skillful feeders have been able to approximate these conditions. The time at which calves can be put on milk substitutes depends upon the same factors as in the use of separated milk, namely, the breed, development and vigor of the calves, etc. It is hardly safe, as a rule, even with the most vigorous ones, to attempt to put them on milk substitutes alone within one week after birth; and with calves below the normal in vigor, some milk for two weeks or more may be necessary to raise them. In supplying a substitute for milk an attempt is usually made to use a liquid, the composition of which resembles milk as much as possible. The following milk substitutes are among those in use:

Bean Soup—Bean soup for calf feeding is prepared as follows: Parboil the beans in soda and drain the water off, add water and boil the beans until soft, then press through a colander. A quarter of a pound of beans in four pints of water constitutes one feed at first. This should be gradually increased until the calf consumes one pound of beans at a feed, on the basis of two feeds a day.

English Preparation No. 1—Wheat flour, 1 lb.; flaxseed meal, 2 lbs.; linseed meal, 1½ lbs. Stir ¼ lb. of the mixture into six pints of boiling water for one feed (twice a day) at first. Gradually increase until the quantity is doubled. (In preparing flaxseed or linseed for calves, it should be boiled with water or very thoroughly scalded. If merely soaked in water (cold or warm) the conditions favor the production of a poison. On the other hand, if it is fed whole or simply crushed, there is no risk of poison).

English Preparation No. 2—Linseed meal, 2 lbs.; oatmeal, 2 lbs.; flaxseed meal, 1 lb. Mix 1 lb. with 7 pints of boiling water and allow to stand overnight. Next morning take one-half of the mixture, add water enough to make 5 pints, boil for 10 minutes and add ¼ teaspoonful of salt and 2 teaspoonfuls of sugar. This makes one feed for the first few days that the calf is put on this ration and fed twice a day. Gradually increase until quantity is doubled.

Hayward's Calf Meal—(As prepared and used by Prof. H. Hayward at the Pennsylvania Agricultural Experiment Station, State College, Pa.):

Flour, 30 lbs.; coconut meal, 25 lbs.; nutrium or dried skimmilk, 20 lbs.; linseed meal, 10 lbs.; dried blood, 2 lbs. One-half pound of this mixture is stirred into 3 pints of boiling water, and when sufficiently cool constitutes one feed when the calf is fed twice a day. This is the ration at the start; the quantity is gradually increased as for the English preparations.

One of the objections to this mixture is that some of the ingredients are not readily obtained in all sections of the country.

Lindsey's Calf Meal—(As prepared and used by Prof. J. B. Lindsey at the Massachusetts Agricultural Experiment Station, Amherst, Mass.):

Ground oat flakes, 22 lbs.; flaxseed meal, 10 lbs.; flour middlings, 5 lbs.; fine corn meal, 11 lbs.; prepared blood flour, 1½ lbs.; salt, ½ lb. The gruel is prepared in the usual way, by adding a little cold water to the dry meal and then about 5 pints of boiling water for each half pound of meal. The mixture should be allowed to stand until cool, and always warmed to 90 degrees or 100 degrees Fahrenheit before feeding. The directions for feeding are the same as for Hayward's meal.

Skimmilk Powders—One pound of powder is mixed with a small quantity of cold water to prevent the formation of lumps, then stirred into 9 lbs. of boiling water and fed in the same proportion as milk.

The quantity of this powder available for calf feeding at a cost within easy reach of the farmer is limited. First-grade skimmilk powders cost too much to feed to calves. In the manufacture of skimmilk powder, however, a limited quantity of low-grade product is made, not good enough to sell for bakers' or confectioners' use, which usually can be purchased at a price that permits its use in calf feeding. Under present conditions it is questionable whether these powders have wide use in calf feeding.

In feeding all milk substitutes the quantity fed should be substantially the same as when separated or whole milk is fed. If, however, there are indications of scours, the quantity should be reduced. The following may serve as a guide in using milk substitutes for feeding strong, vigorous calves:

First week—Whole milk.
Second week—Whole milk.
Third week—Three parts whole milk, one part gruel.
Fourth week—Three parts whole milk, one part gruel.
Fifth week—Whole milk and gruel, equal parts.
Sixth week—Whole milk one part, gruel three parts.
Seventh week—All gruel.

All milk substitutes lack a great deal of being as satisfactory as either whole or skimmilk, and milk has to be very high in price to justify the use of milk substitutes during the first two weeks of the calf's life.

Grain and roughage should be fed with milk substitutes the same as with separated milk.

QUARTERS FOR CALVES Small calves should not be bumped and jostled about. An easy way to prevent this is to provide small pens, not less than 4 by 6 feet in size, in each of which a calf may be kept for the first two weeks. The pens should be fitted with feed boxes for grain and racks for hay. After the calf is old enough to run with the others it is placed with them in a larger pen. Stanchions are fixed on one side of this pen to provide for the separate feeding of the calves, so as to insure that each receives its proper share. Racks for hay should also be placed within easy reach of the calves.

Too much emphasis can not be laid upon the necessity of having light, dry quarters for the calves. Bedding always should be abundant and should be changed often, in order that the pen always may be dry. Lack of attention to these matters is very likely to allow the development of the various calf diseases.

After the calf is a few weeks old, it can stand considerable cold if it is kept dry and has dry quarters. Provision also should be made to allow the calves plenty of exercise. A small paddock or pasture adjoining the calf stables is excellent for this purpose. Except for the very young ones, calves may be let out in the exercise lot for a short period each day when the weather is not too cold or stormy.

STANCHIONS FOR CALVES If the calves are kept together in a large pen it is very difficult to feed them by hand unless they are tied. When they are loose the milk often is spilled, and the larger calves get part of the smaller ones' share. Very simple stanchions may be constructed to prevent losses of milk and insure the equal distribution of the feed. To prevent the calves from sucking one another they should be kept in the stanchions for some time after feeding.

A calf stanchion may be constructed of cheap or scrap lumber. It is usually 36 to 40 inches high and has a 4-inch space for the calf's head.

WATER AND SALT Many feeders fail to realize the importance of providing the young calf with plenty of water. It is a mistake to think that because the calf drinks milk it does not need water. After the calf is two weeks old it should have access to plenty of fresh, clean water at all times, and when it is old enough to eat roughage it should have access to salt.

MARKING CALVES FOR IDENTIFICATION It is important that each calf be marked plainly so as to permit of easy identification. This is particularly necessary in pure bred herds, and should be done in all herds of any considerable size, even if composed of grades.

A number of marking systems have been in general use, among which the following are some of the most common: Leather strap around the neck, with brass tag attached; ear tags of various forms; notches and holes in the ears, and tattoo marks in the ears.

The leather strap with tag around the calf's neck is a convenient method of marking. The advantages of this system are that the number may be observed readily and no disfiguration of the animal is necessary. The cost, however, is somewhat higher than that of other systems, because of the first cost of the straps and tags and because their renewal is sometimes necessary.

Ear tags of various materials and forms are perhaps the most common means of identification; these tags are inexpensive and easy to attach, but have the disadvantage of being easily torn from the ears.

Marking animals by means of holes and notches in the ear is practiced in many herds, and a number of systems are in use for this purpose.

The identification of an animal by means of tattoo marks in the ears has become quite common in recent years. This system of marking animals has the advantage of not disfiguring the animal, and if properly applied the tattoo is permanent, so that the chance of losing the identity of the animal by the loss of the tag is greatly lessened. On the other hand, it requires close inspection to distinguish the marks in the ears.

SCOURS FROM INDIGESTION Diarrhoea, or scours, is probably the most common disease of calves. Great care must be taken at all times to prevent this condition, as it always hinders the growth and development of the animal and in addition is often hard to cure. This disease is the result of disturbance of the digestive apparatus of the calf and may be caused in a number of ways, the more important of which are the following: Irregular feeding, overfeeding, sudden change of feed, fermented feeds, feeding dirty or sour milk or milk of diseased cows, the use of dirty milk pails or feed boxes, and damp dirty stables. As soon as scours is discovered it is best to separate the affected calf from the other and carefully disinfect the pen. The feed should be reduced immediately at least one-half, milk pails cleaned and sterilized, feed boxes cleaned and disinfected, and any other causes mentioned above eliminated.

A number of preparations are used to treat this disease, a few of the more common of which are blood meal, a teaspoonful at a feed; white of egg, limewater, etc. A dose of four drops of formalin to each quart of milk has been used to advantage, and a drench of three ounces of castor oil followed by a teaspoonful of a mixture of one part salol and two of subnitrate of bismuth also is recommended. Ordinary white clay, mixed with water to the consistency of thick cream, and given in doses of quarter or even half pint, three times a day, has been used recently, with excellent effect.

WHITE SCOURS White scours, or infectious dysentery of the calf, generally affects a number of calves in a lot, and first appears shortly after birth as a diarrhoea with light-colored, offensive droppings. During the course of this disease the calf wants to sleep all the time and can not be induced to suck or drink. It is also very much weakened by the disease and usually dies within three or four days. As far as the department knows, there is no specific method of curing the disease. Ordinary white clay, mixed with water to the consistency of thick cream, and given in doses of a quarter or even half a pint, three times a day, has been found to be very valuable. Manufacturers of biological products, however, are now selling a potent serum which they claim to be effective in both prevention and cure.

Prevention consists in the use of sanitary precautions, such as clean, dry and disinfected pens for calving, and careful disinfection of the navel of the calf at birth, painting the cord with tincture of iodin, and tying it with silk thread. As this disease is of so serious a character that it may cause the loss of a season's crop of calves, the details for the control of an outbreak should be referred to the state live stock official or to a qualified veterinarian in the community.

FEEDING—SIX MONTHS TO ONE YEAR OF AGE As has been previously stated, it is a common practice among dairymen to feed skimmilk until the calf is approximately six months of age. Usually the time of weaning depends upon the availability and cost of the milk.

When milk is fed in abundance it furnishes the greater part of the protein necessary for the growth of the animal. (Protein is an important component of animal food; one of its chief constituents is nitrogen. It is necessary for the making of blood, muscle, skin, milk, etc.) If no milk is fed it becomes necessary for the protein to be provided from some other source. Probably this can be done most economically by the use of some legume, such as alfalfa, clover, soy beans or cowpea hay. When hay of this sort is not available it is necessary to provide the bulk of the protein through a grain mixture. In either case, plenty of roughage should be supplied to the growing heifer at all times. During summer, when good pasture is available, the heifer needs no supplementary feed, although a little hay and grain are sometimes advisable late in the season to insure steady growth.

Part of the roughage should be silage, if it is available. A heifer of six months to one year of age will consume from 5 to 15 lbs. of silage a day. The grain mixture used may be the same as mentioned for calf feeding; any one of Nos. 1, 2, 4 and 5, together with all the alfalfa, clover or cowpea hay that the heifer will eat. In case no leguminous hay such as that just mentioned can be obtained, No. 3 is advised, because it contains more protein. Another excellent grain mixture, to be used when such hay is lacking, is composed of two parts of corn meal, two parts of linseed meal and one part of bran.

The quantity of grain to be fed depends very largely upon the individual animal's growth and condition, as well as upon the price of the grain. Some feeders desire a rapid growth of the young animal, and for this reason feed heavily with grain, while others are satisfied with a slow growth and try to carry their young stock largely on roughage. Either extreme is unwise and a medium course between the two is advisable. A safe rule to follow is to feed 1 lb. of grain for the first hundredweight of the heifer and ½ lb. for each additional hundredweight.

ONE TO TWO YEARS OF AGE After the heifer reaches one year of age, the following rations are suggested: Corn meal, fed according to the rule just mentioned, together with all the alfalfa, clover or cowpea hay that the animal will consume. If no leguminous hay is available, grain composed of two parts corn meal, one of bran and one of linseed meal, gluten meal, or cottonseed meal, and 10 to 20 lbs. of silage ,together with all the dry roughage that the animal can consume, will be found to be adequate. Under ordinary circumstances a gain of at least a pound a day from the time of weaning to the time of first calving is a good average for a dairy heifer.

AGE TO BREED Ordinarily it is planned to have the heifer enter the milking herd between 24 and 30 months of age. No arbitrary time can be set, as this depends upon several factors, such as the size and condition of the animal and the breed to which she belongs. Undersized and ill-conditioned animals should be allowed more time to complete their growth and to improve in condition before entering upon the strain of calving and the ensuing lactation period. It is very important that the heifer make a good growth before she is bred, because after that time she will make little body growth until she has completed her first lactation period.

If heifers are bred to a heavy bull, care should be taken to see that they are not permanently injured. Often times a breeding rack is found to be of advantage; such a rack is inexpensive and easily constructed. A better plan is to use a young bull on the heifers, which eliminates danger of injury.

HANDLING YOUNG HEIFERS Young heifers should be handled as much as possible in order that they may not be shy when they enter the herd. A good plan is to bring the "springing" heifer up to the barn some time before she is due to calve and get her accustomed to the halter and stanchion and to being handled. A little care in this way often prevents considerable trouble after calving, and usually insures a gentle cow

FALL CALVING ADVISABLE From the standpoint of the dairyman who raises his calves, fall calving is desirable under most conditions. Under this system the calf receives milk for the first few months of its life, and at the time when it becomes necessary to wean it a succulent feed in the form of pasturage is available. As a result the calf usually makes uninterrupted gains at a minimum cost. Fall calving has the additional advantage that the bulk of the milk is produced at a time when prices are the best and when it is easiest to maintain a steady flow, and the calves are of the right age so that with careful management they may be bred to freshen in the fall or winter for the first time. This is desirable from the point of view of obtaining the longest milking period in the young heifer.

THE YOUNG BULL The bull calf should be separated from the heifers at about four months of age. His treatment and feeding should be identical with that of the heifer except that to get maximum growth he should receive a little larger quantity of grain. If properly handled, the young bull is ready for light service at the age of from ten months to a year. Too much service before he is two years of age will do him permanent injury, which, of course, should be avoided. It is important that he be properly trained to halter, as this will make him much easier to handle when he is old. At six months of age a ring should be put into his nose.

Among some breeders it is the practice to remove the bull's horns at two years of age. It is asserted that this tends to tame him and prevent him from becoming vicious. One thing that should always be kept in mind is that exercise is essential to the proper development of a young bull and to the health and vigor of a mature one. A small paddock, with a shed for protection against stormy and windy weather, will give him room for plenty of exercise and keep him in good condition.

PURE BRED CATTLE IN U. S.

The Department of Commerce, through the Bureau of the Census, announces the following figures from the 1920 census of agriculture for the United States.

The 1,981,514 purebred cattle in the United States on January 1, 1920, according to the Fourteenth Census, included 1,064,912 cattle of beef breeds and 916,602 cattle of dairy breeds.

The 916,602 purebred dairy cattle were distributed according to breed as follows: Ayrshire, 30,494; Brown Swiss, 8,130; Guernsey, 79,455; Holstein-Friesian, 528,621; Jersey, 231,834; and all other dairy breeds, including animals with breed not specified, 38,078.

Of the 231,834 Jerseys reported, 23,842 were in Ohio, 18,718 in Texas, 13,411 in New York, 11,036 in Pennsylvania, and 10,708 in Missouri. This breed is more widely distributed, perhaps, than any of the other breeds. In many of the Southern States, in particular, the number of Jerseys exceeds the number reported for any other one breed of cattle.

Feeding and Developing the Dairy Cow

(Continued from Page 83)

for stimulating the mammary glands. Malted barley is very high is carbonaceous matter and carries a moderate amount of highly digestible fat.

MOLASSES There are two kinds of molasses—cane and beet. Cane molasses has a more pleasant flavor and hence is more palatable than beet molasses, but beet molasses contains a higher per centage of mineral matter and is therefore more laxative. Molasses consists of the settlings of sugar kettles and although it is a very thick, sticky mass, it has a very pleasing taste and greatly increases the palatability of the feed. As a whole it is highly carbonaceous but is valuable in the ration more on account of its mineral content than the amount of sugar which it contains.

LINSEED MEAL Old Process Oil Meal or linseed meal that is cold pressed, leaving many of the mineral constituents and some of the oil in the residue, is the most desirable. Meals resulting from processes in which naptha or other chemicals have been used often contain proteins that have a lower feeding value. This may be due to the fact that some of the mineral constituents have been removed and some of the proteins made less digestive, as with other grains, linseed oil meal should be ground extremely fine. In fact this is more essential in the case of linseed meal than with the feeds.

COTTON SEED MEAL If the seeds are separated from the hull, cotton seed becomes much more valuable as a feed for dairy cows. Often a large portion of the hull is ground with the seed and in the cotton countries where it is fed without other feeds this may be beneficial, for the feed it not as concentrated. However, for high production, hulls are of no value as the amount of fiber is high and the digestibility is low. Good cotton seed meal should be concentrated and contain at least forty-three per cent of protein. Cotton seed meal is one of the richest of all feeds in protein and fat. These nutrients are readily assimilated and are conducive to the production of large quantities of amido acids of such quality that they are readily converted into milk. Cotton seed meal possesses this quality to a greater extent than any other feed.

FLAXSEED When cows are fed large quantities of concentrated feeds there is frequently a tendency toward constipation which greatly affects the fat content of the milk. Flaxseed, which is extremely high in oil, is very beneficial, and when cows are fed to the maximum, often has a slight influence on the fat content of the milk. Flaxseed is also beneficial on account of the fact that it softens and loosens the skin of the animal. Flaxseed should be used only in limited quantities, as large amounts are extremely laxative.

PEANUT KERNEL MEAL Peanuts are exceptionally high in protein and oil. The oil is quite digestible which makes it very desirable in a dairy ration. This does not refer to the peanut meal which is commonly on the market, but rather to peanuts that are hulled, cracked and ground in a special grinder which produces a hard, dry powder. When ground in an ordinary grinder, peanut paste or butter is the result. Since pure peanut meal quickly becomes rancid, the peanuts should be ground frequently and in small quantities. However, if mixed with other less oily feeds it will keep a longer time without becoming rancid. As is the case with flaxseed meal for high producing cows, peanut meal assists in fat production when cows are fed to the maximum.

SOY BEAN MEAL Soy beans likewise are rich in oil and protein, and the oil being of a different composition, assists in assimilating the other oils. The oil is not as valuable as a feed as peanut oil. A part of the fat has been extracted in ordinary commercial soy bean meal. The fat in soy bean meal is less digestible than in many other feeds and for this reason it is not as important in the ration.

SUNFLOWER SEED Sunflower seed is very rich in oil protein, the oil being of assistance in digesting the other oils. It is therefore beneficial in the production of butter fat.

MINERAL RATION Salt is a necessary part of all rations. It has a physiological effect upon the system, stimulates the appetite, increases the palatability of the feed, stimulates the secretion of the digestive fluids and hastens the circulation of the blood and secretory fluids. It also assists in preventing undesirable fermentations.

It is extremely desirable that a cow producing large quantities of milk should drink a large amount of water, and salt develops thirst. The amount of salt to be given should be governed by the amount of milk that the cow gives and also by the character of the feed that she eats. A cow weighing one thousand pounds and giving as much as 16 lbs. of milk a day, should never have less than one ounce of salt per

(Continued on Page 91)

Feeding and Developing the Dairy Cow
(Continued from Page 90)

day. A high producing cow should receive from two to three ounces.

BONE MEAL OR CALCIUM PHOSPHATE Other minerals are quite essential in the ration when a cow is giving a large amount of milk, as milk production causes a steady drain upon the phosphorus and lime especially, in the cow's system. It was formerly believed that ground rock phosphate would supply these nutrients, but experience proves that it does not serve the purpose as well as the phosphorus derived from bone.

Another method of increasing the calcium phosphate is to feed plants grown on soil that is rich in this element. However, this does not ordinarily supply it in sufficient quantities for the high producing cow so the addition of a little bone calcium phosphate is very beneficial in a ration. In no case should raw bone meal be fed unless it has been properly sterilized and finely ground.

BLOOD MEAL Blood meal offers one of the greatest sources of minerals, proteins and vitamines that is found in any feed. It is very easily assimilated but the average cow will not eat it unless the odor is covered with condiments and the flavor changed. For this reason blood meal must be mixed with some highly aromatic compound so that the palatability of the ration will be maintained. Three to six ounces of blood meal per day will make a splendid addition to the high test ration. Blood meal must be properly steamed and sterilized and kept perfectly dry, for if kept in a moist place it becomes dangerous as a feed.

IRON SULPHATE Iron sulphate acts as a preventive of undesirable fermentations. For the ordinary cow, salt answers this purpose but the quantity given is not proportion to the needs of the system of high producing cows. Not more than four grams of iron sulphate should be given in the daily ration.

CARLSBAD SALTS Carlsbad salts act as a laxative and increases the peristaltic action to some extent. They assist in keeping the fecal matter in a more plastic condition so that it is more readily expelled.

CHARCOAL Charcoal may be given to the cow whenever there are signs of gastric fermentation. This is indicated by the way a cow belches and also by the production of the intestinal tract.

DEFICIENCIES Deficiencies consist of fenugroek, millet seed, mustard seed, juniper berries, fennel, anise seed, caraway seed. These contain various elements in such form that they furnish a general tonic to the whole system, supplying the vitamines and a large number of unknown nutrients which are, as a matter of fact, the limiting factors in digestion. The seeds, especially millet seed, must be finely ground.

STIMULANTS It is not permissible to give stimulants of any kind when cows are on official test and there is probably nothing that will satisfactorily increase the flow of milk and the amount of butter fat like good ration. In fact milk itself is probably more conducive to milk production than any other feed. Stimulants will not increase the flow of milk permanently and drugs should be given only in special cases and for a short time, for the purpose of aiding the digestive organs and renovating the cow's system. If the animal seems sluggish it is evident that the system has been clogged. This may be likened to clinkers that accumulate in a grate, or soot in a furnace.

Under these conditions it is well to eliminate one of two feeds. Give one pound of Carlsbad salts, one ounce of iron sulphate and one ounce of gentian. Sometimes it is more effective if .08 of a gram of nux vomica is added. Following this there might be given as a stimulant four grams of aloes, .08 of a gram of nux vomica and six grains of gentian. If the cow does not fully recover her appetite after this treatment, fifteen drops of Fowler's solution of arsenic may be given each day for a week. It is probably better to give ten drops the first day and increase the dose until she receives twenty drops on the fourth day, then gradually decrease the amount.

If the cow eats but does not produce, try drenching her with the following mixture: 10 per cent grain alcohol, 10 per cent syrup, 20 per cent oil and 60 per cent water. A six per cent solution of alcohol and malt extract poured over the grain is another good stimulant. If the cow does not then produce try next ten to twenty five grams of black sulphide of antimony and from two to five of sulphur praeciptitatum. To stimulate the secretory organs give ten to twenty grams of jaborandi for ten days, or from six to eight grams of the extract of pilocarpine for six days. A dose of digitalis will increase the circulation of the blood. When the cow has regained her normal flow of milk, regular feeding operations should be resumed. Great care should be taken that stimulants are not given in too large amounts or over too long a period of time, as more harm may be done than good accomplished.

(Continued on Page 92)

Feeding and Developing the Dairy Cow

HAY AND ROUGHAGES Of the hay and other roughages probably alfalfa hay is the best. Sweet clover, if cut early and properly cured, comes next, and clover hay after this. Seventy-five stems of the hay, hence the shatterings or per cent of the protein is in the leaves and the leaves of the hay are the most desirable part. If the hay is finely cut or preferably ground, the digestibility is greatly increased but the cost of grinding is an objection to this treatment of hay. Other hays, like timothy, should not be fed in large quantities, however, when mixed with clover, timothy sometimes becomes desirable on account of the fact that is acquires some of the protein from the clover. Oat straw should not be consumed by the cow in large quantities, but a little piled up at the side of the stall and later used for bedding is not undesirable. A high producing cow should not have much straw for the fiber content is very high and the digestible nutrients low.

PASTURE If the grass is available it is desirable to replace part of the other roughage with grass. It is best to gather it, or a cow may be tethered for a time each day in a comfortable place. Pasture grass usually reduces the flow of milk in the beginning but increases it in the course of a few days. No explanation can be made of the stimulating effect of grass other than it contains vitamines and other substances which play an important part in metabolism of the cow. Blue grass and white clover are preferable to other pasture grasses.

WATER It is highly important that at all times the cow be supplied with good fresh water at the proper temperature. Water at a temperature of from 55 to 60 degrees, both winter and summer, is the most desirable.

STEAMING To insure perfect freedom from all fermentations the feed may be steamed by inserting a steam pipe into the bottom of the bin, or better by pouring boiling water over the feed when it is in the individual pails and covering it to keep it hot. This method makes the feed less palatable in the summer but increases the palatability in the winter. When there is the slightest indication of mold in a feed, no chances should be taken in feeding it to a high producing cow. As a matter of fact moldy feed should never be fed to any cow. Steam is a sterilizer and thoroughly steaming moldy feed makes it fit for consumption.

QUANTITY OF FEED TO BE FED In the case of high production the quantity of feed to be fed must not be determined by the quantity of milk that the cow produces, but rather by her ability to consume. If, in the course of about three months, she does not respond to feeding, she is not likely to be a cow with a capacity for high production. Feed should be weighed so that the exact quantity required for each feed will be known. The amount should be increased gradually at the rate of a pound per day. One of the tricks of good feeding is to get the cow to consume large quantities of feed. Make the feed as palatable and appetizing as possible and make all conditions surrounding the feeding as pleasant as possible, so that she will constantly increase her consumption. Cows must be gradually trained to consume large quantities of feed. Sometimes as much as a year is required before a cow can consume regularly twenty-four pounds of feed, when she had been in the habit of consuming but eight or ten pounds. Some cows will eat as much as twenty-four pounds of feed a day. Naturally, in order to do this she must be trained to consume this large amount.

Great care should be taken to gradually increase the amount fed, for when increased too rapidly, the quantity of digestive fluids is not sufficient to take care of it, the feeds ferment in the intestinal tract, toxins develop and indigestion is the result. Whenever the fecal matter becomes more or less dark and firm, a dose of salts or oil should be administered. In the case of indigestion it is wise to stop feeding for a few rations, gradually starting out again with beet pulp, bran and about a pound of Glauber salts. A cow in good condition should recuperate in a few days.

MASSAGING Before parturition the udder should be gently massaged by raising it against the abdomen and moving it from side to side with a twisting motion and gently rubbing it downward. Ths should be done at least three times a day. After parturition the udder should be massaged after each milking. This relieves the swelling to a great extent.

MILKING If a cow weighing a thousand pounds gives more than 18 lbs. to a milking, she should be milked three times a day. A larger cow may conveniently retain a larger amount, but any cow giving more than 45 lbs. of milk a day, should be milked four times. Increasing the number of milkings, always has a tendency to increase the flow of milk. It is, however, not practical except in extreme cases to milk more than four times and is not per-

Feeding and Developing the Dairy Cow

mitted when cows are on official test. Cows should be milked regularly, quite rapidly, and thoroughly dry at each milking. Most cows give more milk when milked by hand, especially if they have been trained to hand milking; however, if trained to be milked with a milking machine, they will often respond better to the machine than to hand milking and should then be milked with the machine even when on test. When a milking machine is used, care must be taken that the rubbers maintain their resiliency and whenever they lose their elasticity, new rubbers should replace the old.

Milking should always take place when the cow is being fed. Eating stimulates the secretory glands and if the animal is allowed to eat in comfort and does not have to hasten for fear she will lose some of her feed by other cows taking it away from her, the flow of milk will increase materially. A cow giving a large flow of milk must necessarily have a large sphincter muscle at the end of the teat in order to hold the milk in the milk cistern. If the end of the teat is not properly taken care of by using a little vaseline to soften the sphincter muscle, the operation of milking will hurt the cow and consequently cause her to hold up a part of her milk. This is especially true with high producing cows. Milking should be done in as easy a manner as possible. Hard squeezing of the teat or excessive pulling interferes with the flow of milk. However, a small amount of gentle massaging with upward movements of the teat are necessary. The milker's judgment of the condition of the cow, will to a great extent govern this factor.

COMFORT OF THE COW A cow producing a large volume of milk should never be kept in a stanchion, for all the muscles of the body can not be comfortable nor as healthy as when kept in a box stall.

Previous to parturition a cow should be placed in a box stall which has been cleaned, disinfected and whitewashed. In no case should white lead and oil be used about the stall as this frequently causes lead poisoning. It is also dangerous to use lead paints in the mangers. It is not best to have a cement floor in the stall. A cork, creosote block or wood floor is much more satisfactory. It is also advisable to have the mangers made of some other material than concrete for licking the cement irritates the cow's tongue and a smooth surface such as wood or galvanized iron is preferable. The best way of feeding a cow is from individual galvanized iron pails or small individual wood boxes that can be properly washed each day.

A box stall should be so arranged that there is plenty of fresh air and light. It is much better to have the temperature lower and keep the cow blanketed, than to maintain a higher temperature. Sixty to 65 degrees in a clean, dry, well bedded stall, is the most desirable temperature. Blanketing stimulates the oil glands of the skin and likewise helps to stimulate the oil glands of the udder. During July and August when the temperature is high, cows producing large quantities of milk are under a great strain and if forced too heavily, are likely to suffer from nervous exhaustion. This is one of the chief reasons why cows in northern climates can be made to work harder than those where the climate is warmer.

The most difficult time of the year to make a record is in July and August. In extremely warm weather feed is much more likely to ferment and often starts fermentation in the alimentary canal which results in a profuse diarrhoea. Feed should be kept very dry during the summer months.

A cool, well insulated stall should be provided for the cow. There should be double windows and some method of darkening it during the day to prevent annoyance from flies. An electric fan is absolutely necessary when maximum production is desired. This may seem an extravagance, but at the ordinary rate paid for electric current the cost is usually less than the price received for the additional amount of milk.

Whether or not electricity is available, there should be provided a ventilating system that will temper the air. The air may be drawn in through tile laid about five or six feet in the ground. They should be about three inches in diameter and not less than six tile rows per cow should be laid for a distance of about 200 feet. These should be drained to a low point so that no water will stand in them and the air can circulate freely. A cistern at the end properly cemented, with a cowl to force the air into the barn is good wherever no electricity is available. Wherever electricity is available the air may be forced in with a small fan. Another method which seems to work satisfactorily is where the air circulates through saturated cloths. The evaporation absorbs the heat, cools the air and lowers the temperature considerably. Low temperature is conducive to heavy work during the hot weather.

RUBBING AND MASSAGING THE SKIN A cow that is blanketed should be curried and massaged every day. After the skin has been properly curried with a rough curry comb and brushed with a stiff brush, pick up folds of skin and pull them back and forth. This is especially desirable over the ribs and over

Hamilton County Stock Farm

LEWIS F. MARTIN, Prop.

NEWTOWN - - OHIO

(Hamilton County)

Breeder of
HIGH CLASS JERSEYS
Herd Sires:

Oxford Daisy's Dandy Fox 140725

Sire—Oxford Daisy's Flying Fox, sire of 38. Dam—Imp. Dandy Baroness, R. of M., 10,474.5 lbs. milk, 553 lbs. butter.

You'll Do's Jap 194685

Has 2 Gold Medal Grandsires; Two Gold Medal Sisters.

Sire—Heir of Crystal Spring. Dam—You'll Do Success, R. of M., 12,121 lbs. milk, 780 lbs. butter.

Young Bulls by these sires from high testing dams for sale at moderate prices.

Also O. I. C. and Chester White Swine; Standard bred Single Comb Brown Leghorn Poultry.

Lewis F. Martin, Newtown, Ohio

ALFA MEADE FARM

MAJESTY JERSEY CATTLE

EARL SMITH & SON, Owners.

Westville, Ohio

PRODUCERS THAT REPRODUCE

Feeding and Developing the Dairy Cow
(Continued from Page 93)

the backbone. Every week certain portions of the cow's body, especially the backbone, should be washed with warm water and castile soap. After drying thoroughly rub the skin with a solution of alcohol and peanut or linseed meal, equal parts.

EXCRETA The excreta is the indicator of the cows digestion. When this becomes dark and when discharged contains the peristaltic folds, it is evidence of constipation. The excreta should at all times be of a brownish, plastic consistency and contain no undigested material.

P. S. This article originally written by Prof. Oscar Erf for "The Jersey Bulletin," it being one of the best articles the writer has ever read on the subject. It is published herein by special permission.—Editor.

CONSTITUTION OF THE OHIO JERSEY CATTLE CLUB

ARTICLE I—NAME

Sec. 1—The name of this organization shall be The Ohio Jersey Cattle Club.

ARTICLE II—MEMBERSHIP

Sec. 1—Any person interested in Jersey Cattle may become a member upon payment of the annual dues, namely five ($5) dollars. Juvenile membership is one ($1) dollar annually.

ARTICLE III—OFFICERS

Sec. 1—The officers shall consist of a President, Vice President and Secretary-Treasurer.

Sec. 2—The duties of the officers shall be such as usually pertain to these offices.

ARTICLE IV—BOARD OF DIRECTORS

Sec. 1—The Board of Directors shall consist of the officers and one director from each county club, and they shall continue in office for a period of three years. However, until a balanced program of appointment shall have taken place, a certain portion of said directors shall be appointed for a period of two or three years each, after which at each annual meeting one-third of the directors will be elected annually.

Sec. 2—The duties of the Executive Committee shall be to plan all work and have charge of the disbursements.

ARTICLE V—MEETINGS

Sec. 1—The annual meeting shall be held during Farmers' Week at the Ohio State University.

Sec. 2—Special meetings may be called by the President or Secretary.

ARTICLE VI—QRORUM

Sec. 1—Twelve members shall constitute a quorum at any meeting.

ARTICLE VII—AMENDMENT

Sec. 1—This Constitution may be amended by two-thirds vote of the members present at any meeting called for the purpose.

THE OHIO JERSEY — OHIO — NINETEEN TWENTY-TWO

Official List of Accredited Jersey Herds in Ohio

As furnished by
MR. B. H. EDGINGTON, State Veterinarian

Name	No.
Geo. Abbott, Chippewa Lake	23
Charles Alexander, Lisbon	28
C. W. Allcorn, Bellaire	9
F. L. Allyn, Port Clinton	3
Homer P. Alwine, Quaker City	13
I. J. Ambler, Jacobsburg R. 2	12
Arbaugh & Meiser, Salem	11
W. L. Bachman, Canal Winchester	21
H. C. Bachman, Canal Winchester	17
E. D. Bailey, Barnesville	29
L. P. Bailey & Son, Tacoma	99
O. J. Bailey, Tacoma	36
Ross Bailey, Barnesville	20
W. H. Baker, Salem	13
H. S. Bartlos, New Philadelphia	11
H. J. Beardsley, Canfield	50
W. L. Beardsley & Son, Ellsworth Station	24
Walter Bee, Bethel	21
V. F. Beeghley, Dayton R. 4	6
Belmont Children's Home, Tacoma	11
M. E. Beman, Thurman	26
S. H. Bennett, Salem	27
Peter Beninghofen, Hamilton	10
John A. Binns, Salem R. 5	24
Black Bros., Powell	49
W. A. Black, Cadiz	36
Daniel Blackburn, New Waterford	5
Wm. J. Blackburn, Salem	53
A. H. Boltz, New Philadelphia	13
Hugh W. Bonnell, Youngstown	62
S. B. Bowles, Harrison	10
Bowling Green Normal College, Bowling Green	8
C. Raymond Boyer, Montpelier	13
J. C. Boyer, Montpelier	8
W. G. Brady, Barnesville	12
J. & J. C. Brantingham, Winona	60
R. O. Breese, Ohio City	4
Brooks & Barker, Salem	42
Charles E. Broome & Son, Coshocton	17
Walter E. Brown, Hubbard	20
Fred Bruderly, Washingtonville	21
Frank Buchwalter, W. Austintown	17
D. C. Bundy, Barnesville	5
George H. Burchfield, Barnesville	13
T. L. Calvert, Selma	22
Henry Candel, Columbiana	19
C. Cal Carter, Wapakoneta	5
B. H. Chidlaw, Cleves	25
Grant G. Chidsey, Brunswick	31
C. C. Cope & Son, New Waterford	16
Robert Copeland, St. Clairsville	17
R. W. Coppock, Winona	10
Rufus Cox, Manchester	24
Thomas Crawford, Salem	7
James W. Creeger, Tiffin	15
C. C. Creek, Montpelier	39
M. L. Cunard, St. Clairsville	18
L. S. Daley, St. Clairsville	27
C. W. Damon & Son, Brunswick	23
H. B. Dargitz, Montpelier	13
Solomon Deitz, Millersburg	11
G. W. Delaney, Barnesville	12
L. L. Denison & Sons, Delaware	36
Lewis Denkhaus, Lisbon	19
Chas. W. Deweese, Salem	10
George Dimmitt & O. E. Schooley, Newtonsville	9
J. E. Doudna, Quaker City	22
Doudna Sisters, Barnesville	9
G. E. Dudley, Youngstown	11
I. N. Dumford & Son, Newtonsville	9
J. W. Dumford, Pleasant Plains	11
R. L. Dunipace, Bowling Green	15
Walter Dunn, Salem R. 2	26
W. H. Dysart, Pataskala	16
W. G. Edgerton, Hanoverton	16
Ray Elliott, New Concord	11
G. B. Ellsworth, Bowling Green	8
D. C. Emery, Napoleon	13
Harry Erlenbach, Sta. E, Columbus	18
L. H. Everett, Lisbon R. 3	11
Jay Farber, E. Sparta	11
Herbert Farrell, Sandusky	7
Frank Filbrun, Dayton R. 4	9
Geo. E. Fisher & Son, Westerville	14
John Fitz, Venice	30
Wm. Foreman, Jacobsburg	4
Fred Foster, Gallipolis	35
Homer J. Frank, Bowling Green	18
Allen Frederick, Poland	32
R. E. Frederick, Poland	30
Homer French, Salem	17
Friend's Boarding School, Barnesville	29
U. W. Garber, Dayton R. 4	2
H. B. & B. F. Garman, Everett	25
Charles Garrigues, Salem	13
W. W. Garrison, Springfield	23
E. E. Gemberling, Kent	14
Lavina Gibbons, Barnesville	3
F. M. Giffin, Bellaire	16
J. A. Giffin & Son, Bellaire	24
O. L. Godfrey, Bridgeport	18
J. M. Graham, Millersburg	12
H. H. Green & Son, Johnstown	18
W. E. Grim & Sons, Columbiana	18
T. J. Grogan, Wilmington	23
Haines & Cochran, Blanchester	78
J. R. Haines, Dillonvale	10
Clifton P. Hall, Salem	20
Herbert D. Hall, New Waterford	9
H. S. Hall, New Waterford	3
J. Wilmer Hall, Barnesville	10
Salmon P. Halle, Wickliffe	31
A. C. Halverstadt, Columbiana	10
Clark J. Halverstadt, Leetonia	10
S. H. Hamilton, Hillsboro	16
M. R. Hanna, Elkton	17
Elias Hartley, Barnesville	5
B. W. Havens, Galena R. 1	24
Albert Hester, New Lebanon	12
Sylvanus Heintzelman, Canfield	5
E. E. Heistand, Dayton	5
James Henderson, Barnesville	22
David Hess, Westerville	27
Caleb Hobbs, Barnesville	23
O. W. Hoffhiens, Bowling Green	10
Harry Holmes, Gambier	6
Carl R. Hoopes, Salem R. 2	10
Chas. B. Hoopes, Salem	16
Joshua W. Hoopes, Salem	3
Elmer L. Horn, New Philadelphia	3
H. W. Ingersoll, Elyria	34
Dawson Irey, Lisbon	9
A. B. Jeffers, Barnesville	12
Pursell Jenkins, Frankfort	15
W. S. Johnson, St. Clairsville	12
A. C. Jones, Yorkville	14
Austin Jones, Wilmington	18
W. L. Jones, Twinsburg	24
James A. Judkins, Barnesville	11
J. W. Kennedy, St. Clairsville	7
W. F. Kennedy, Blue Ash	19

(Continued on Page 96)

COME TO OHIO FOR GOOD JERSEYS —— THREE THOUSAND BREEDERS WAITING TO SERVE YOU.

ACCREDITED JERSEY HERDS
(Continued from Page 95)

Name	Count
Oliver E. King, Tiffin	14
Walter F. Kirk, Port Clinton	16
S. L. Kirkland, Key	13
T. C. Kirkland, Key	5
E. J. Krieger, Swanton	7
Geo. E. Kryder, McClure	21
J. H. Law, New Philadelphia	2
Charlie LeGalley, Bowling Green, R. 2, Box 62	15
Aaron Lentz, Dayton	10
Torrence Lesher, Vandalia	14
H. H. Lewis, Barnesville	10
W. D. Linn, Millersburg	12
Chas. Livezey, Barnesville	2
A. M. Lodge, St. Clairsville	16
C. A. Long, Newtonsville	6
George A. Long, Holgate	10
Geo. Longworth, Felicity	20
Mrs. W. I. Lucius, Hamilton R. 2	14
C. R. Luyster, Key	7
Geo. M. McBurney & Sons, Barnesville	31
D. B. McCune, Salem	6
J. H. McFarland, St. Clairsville	9
B. A. McKitrick, W. Mansfield	14
D. D. McLellan, Gallipolis	5
M. B. McNutt, Wooster	18
W. E. Marckel, Continental	12
S. E. Martin, New Concord	40
Wilson J. Martin, Millport	8
J. F. Martti, Zanesville	30
Edward L. Mead, Bethesda	24
F. M. Meloy, Delaware	8
L. L. Mercer, St. Clairsville	16
C. F. Meyer & Son, New Philadelphia	31
Joseph E. Meyers, Barnesville	19
J. W. Meyers, New Philadelphia	5
D. C. Miley, Layland	12
E. F. Miller & Son, Vermillon	12
H. P. Miller & Son, Sunbury	37
M. M. Miller, Kinsman	25
William Miller, Belmont	24
C. S. Moore, Athens	18
John H. Moore, Pataskala	18
Wm. T. Morr & Son, Wooster	7
A. A. Mullenix, Jamestown	4
D. A. Murphy, Speidel	37
J. S. Neer & Son, Mechanicsburg	35
A. I. Negus, Bridgeport	24
A. J. Nickols, Berlin Heights	26
George Nuhfer, Woodville	14
James R. Orr, Cedarville	9
Charles Ost, Dover	5
Norman B. Patterson, E. Palestine	40
A. I. Patton, Salem	25
A. L. Pemberton, Salem	18
R. J. Perry, McClure	10
C. G. Phelps, Rock Creek	7
C. M. Philips, Barnesville	9
Ross Phillips, Barnesville	12
Clyde Pickering, St. Clairsville	21
G. W. Pike, New Waterford	10
J. G. Pim, Beloit	27
C. H. Plumly, Barnesville	51
Jesse C. Pottarf, Salem R. 3	20
F. Everson Powell, Columbus	11
Lawrence J. Powers, Huron	14
Doyt Price, Van Wert, R. 7	4
W. H. Price, Woodville	15
Myrtle E. & Earl F. Plye, Clarksville	79
P. L. Renninger, Clinton	9
H. F. Richards, Salem	40
John L. Riley, Canfield	14
Starling L. Roads, Blanchester	4
Homer E. Roberts, Mechanicsburg	60
Robinson Bros. & Clark, Plain City	57
E. C. Robinson, Copley	17
Roscoe M. Rogers, Rogers	8
Bert Rose, Greenfield	9
C. W. Ross, Felicity	9
Chas. Rossolot, Pleasant Plains	20
D. C. Roudebush, Newtonsville	7
Harry Roudebush, Harrison	20
Benj. Rupert & Son, New Waterford	25
J. S. Rupert, Columbiana	26
Walter W. Sale, Columbus	7
W. H. Sawyer, Marble Cliff	6
E. C. Sayre, Dover	14
Wm. Schaefer, Bellaire	9
C. V. Schindler, New Philadelphia	8
Grover E. Schneider, N. Canton	12
Dr. J. H. Schnurrenberger, W. Austintown	22
A. J. Schrock, Orrville	11
Fred W. Schwartz, Port Clinton	6
Seth P. Scott, Lisbon	23
E. W. Sears, Barnesville	2
W. H. Sears, Barnesville	6
Karl Shaffer, North Industry	9
C. M. Sheeley, Columbiana	18
Sheffield Farm, Glendale	44
R. F. Shepherd, Hamilton	22
Shipman & Wharton, Alledonia	33
J. D. Short, Xenia	26
C. A. Shuler, Hamilton	20
J. L. Sicker & Son, Coshocton	45
Peter Siegman, Columbus Sta. E. R. 8	51
Homer B. Slagle, Poland	30
Joe Slaughter, Coshocton	11
Bert Smith's Sons, Delaware, R. 3	48
C. E. Smith, Greenfield	10
Earl Smith & Son, Westville	30
Smith & Preston, Conotton	27
Wilbur C. Sommers, Fremont	20
W. S. Souders, Rogers	12
A. E. Stanton, Barnesville	12
Charles M. Stemen, Johnstown	10
J. S. Sterling, Dover	4
W. E. Stewart & Son, Salem R. 5	19
Chas. Stockslager, Lewisburg	23
Thos. G. Stratford, Canfield	17
George Sumption, Centerville	18
D. S. Thompson, Bridgeport	14
W. W. Thornton, Akron	26
Grant E. Tilletson, Brunswick	13
Trimble Bros., Key	18
S. F. Trimble, Key	3
S. P. Troendly, Stone Creek	5
R. B. Troyer, Continental	18
Wilbert Turnbull, Gahanna	21
Chas. Tyson, Hamilton	23
Bert Vincent, Salem R. 3	8
T. E. Votaw & L. H., Salem R. 2	12
J. F. Walker, Gambier	52
R. L. Walker, Bridgeport	25
Sylvanus T. Walter, New Philadelphia	1
W. L. Ware, Batavia R. 7	6
H. S. Warner, Greenville	12
Guy Whinery, Salem	22
Willis Whinery, Salem	85
Jacob E. White, Greenfield	183
L. A. White, Clyde	13
Thos. P. White, Hooker	27
W. A. Wildman, Bridgeport	3
Erwin L. Wilkinson, Brunswick	24
Edgar E. Wise, Bellaire	16
I. B. Witmer, Leetonia	9
Thurman S. Woodward, Youngstown	16
J. C. Zedaker, Youngstown	8
William Zumbrum, Trotwood	9

F. S. Reynolds' herd, of Dayton, Ohio, passed 2 successful tests. Third test will be made this fall.

RICHEST IN SOLIDS OTHER THAN FAT

Jersey skim-milk is also highest in solids not fat, as compared with skim-milk of other breeds.

The United States Government Animal Industry Bulletin 155 gives these figures: Jersey milk, 9.16 per cent solids-not-fat; Ayrshire milk, 8.75 per cent solids-not-fat; Holstein milk, 8.27 per cent solids-not-fat.

This information is valuable in that it shows the superiority of Jersey milk for manufacturing purposes, and in nutritive value.

COME TO OHIO FOR GOOD JERSEYS —— THREE THOUSAND BREEDERS WAITING TO SERVE YOU.

MEMORABILIA

By COL. D. L. PERRY, Columbus, O.

I have just terminated my forty-fifth year as a Jersey cattle auctioneer; forty-five years replete with strenuous activity, happy years, fruitful and rewarding, circumscribed by the joy of living. During four decades and a half of selling I have traversed our continent many times over. I have sold for the good man and the bad man, for the man of wealth and the man of poverty; sold in winter and in summer, in heat and in cold and in sunshine and in rain. And as for the proverbial "gamut of emotion" if ever a man ran it I have run it. I have, perforce, encountered human nature at its best, and at its worst, and out of the wealth of my experiences I could conjure forth many interesting tales with which to beguile your ears. But space is not permitted me here to do so, and, moreover, this task I am reserving for my autobiography which ere long I contemplate writing.

In this article, however, my interest is less in self. This is to be no eulogy in the first person singular, but rather it is dedicated to that other man without whom success for me could never have been; without whose solicitude, co-operation and kindly fellowship I could never have attained any degree of achievement whatever. Him I fondly term "the other fellow," he who has been the most potent factor in what little fame I may have achieved and who has as much made me as I have made myself. The OTHER FELLOW! Have you never considered what proportion of your success should rightfully be ascribed to his unselfish co-operation? Have you never contemplated how vain would have been all endeavor, how meager the fruits of years of labor, how sordid, and how devoid of real, human interest would have been your life minus his benevolent sympathy? Then lend ear when I tell you that he is the main spring of the watch of life, the keynote of your existence, part and parcel of your life.

As for myself, I have slowed down, and, though I am naturally loath to concede the fact, advancing years are bearing to me the lassitude that all men experience when their period of endeavor approaches the half-century mark. And, I reiterate, although I am reluctant to decrease active participation in the game to which my heart strings have been tied for so long a time, yet I am willing to make room for the fellow who is to fill the shoes I shall ere long have vacated. Just now I only make room, against the time when I shall be permitted to make way. I say "permitted" because "the other fellow" still clamors that he will not have me desert him. His letters for dates on each succeeding day, asseverate that he needs me, that he will have no substitute while I am yet capable of withstanding the rigors and hardships of the life. The gratitude I bear him countenances but one answer, "Just as long as you want me, and I am able, I shall do my best for you."

Meanwhile, as the range of my activity decreases, and as I "retire gradually" (as I like to express it), my increasing hours of leisure permit time for many hours of happy retrospection. Now, as I write, I sit on the front porch of my farm, whither I fly after every road trip. I gaze about me and my eye envisions my fields of corn, my outbuildings, my orchard, and my flowers lining the driveway. A feeling of contentment steals over me and gives impetus to kindly philosophications. But chiefly the thought comes to me that what ever measure of achievement has been mine, whatever degree of success I have attained, and if mine has been a life of purpose and accomplishment, it is due far less to my own individual effort or merit than to the beneficent and benevolent interest of my friend, "the other fellow."

Columbus, Ohio, August 24, 1922.

JERSEY MILK—THE KEY TO DAIRY PROFITS.

The creamy richness of "Jersey" milk, the superb quality of "Jersey" cream, the fine flavor and texture of "Jersey" butter are recognized everywhere. Your best card of introduction, your best recommendation, need only be the word "Jersey" on your caps, labels, or containers.

When you have said that your products are the products of "Jersey cows," you have said it all.

You will always have a better market, with better prices, as long as you sell "Jersey" products.

Creameries and milk stations everywhere are paying a premium for extra butterfat content.

Jersey milk is the richest of all breeds and is therefore the key to greater profits from your dairy.

Headquarters for
The Ohio Jersey Cattle Club
AT

THE CHITTENDEN

Columbus, Ohio

—::—

The Chittenden is a hotel established on the principles of courtesy and service.

The food served in our Restaurant and Coffee Shop is the best that can be procured and is prepared to please the most critical taste. The charges are consistently moderate.

N. A. COURT, Manager.

RICHLAND COUNTY JERSEY CATTLE CLUB

IRA E. SMITH,
President,
Ontario, Ohio.

J. W. CLINE,
Secretary,
Mansfield, Ohio.

This organization is making every possible effort to have every Jersey herd in Richland County a clean herd, to have every Jersey herd conducting Register of Merit tests and to develop the many great dairy cows in the county.

When in Need of Good Jerseys Consider Richland County

MAKING BUTTER ON THE FARM
(Continued from Page 44)

drawn from each of three cows, place in the jars, cover, cool to 75 degrees Fahrenheit and keep at that temperature until curdling occurs.

3. Curdling, or coagulation, should take place in about twenty-four hours. An ideal curd should be firm, smooth, marblelike, free from holes or gas bubbles, and should show little or no separation of the whey. It should have a clean, sharp, sour or acid flavor.

4. Select the sample that most closely meets those conditions and propagate it, discarding the others. The selected sample is propagated as follows:

(a) Clean thoroughly and boil for five minutes a quart jar, the top and a teaspoon.

(b) Fill the jar with freshly drawn milk, cover loosely, heat slowly to boiling and pasteurize by boiling gently for thirty minutes.

(c) Cool the milk to 75 degrees Fahrenheit and add a teaspoonful of curdled milk described in section 3 and set away to curdle at that temperature

(d) Propagate the starter from day to day in the same manner described in (a), (b) and (c). The starter described in (c) is the one to use for ripening the cream, and should be added in such quantities as to be one-tenth to one-fifth of the cream to be churned. Starter is put into the cream while the latter is being warmed to the ripening temperature. The ripening process with starter is exactly the same as natural souring except that it takes place in a shorter time.

The desirable temperature at which to churn is that which makes the butter granules firm without being hard. This is usually obtained under normal conditions when the churning occupies thirty or forty minutes. The churning temperature necessary depends upon the season of the year and certain other factors, but is usually from 52 degrees to 60 degrees Fahrenheit in the summer and from 58 degrees to 66 degrees Fahrenheit in the winter. If the cream is churned at 62 degrees Fahrenheit in winter, and the butter comes in thirty-five minutes, with the granules firm, it will be noticed as summer approaches and the cows are turned out to pasture, that the cream churns more quickly and the butter is softer. This is an indication that a lower churning temperature should be used, and thus from season to season the churning temperature is regulated so that the butter granules may have the proper firmness.

When the temperature is either too low or too high, undesirable results are obtained. A low temperature prolongs the churning period unnecessarily, and may even make it impossible to churn butter. It causes the granules, especially when the cream is thin, to form in tiny pellets, like fine shot, many of which run out with the buttermilk. The working of the butter and the incorporation of the salt are accomplished only with great difficulty, and the body of the butter is liable to be brittle and tallowy. Adding hot water to cream to warm it, and using wash water more than 3 degrees warmer than the butter in order to soften it, are bad practices, since they injure the quality of the butter. If the proper churning temperature is used, the butter granules will be of the proper firmness.

Too high a churning temperature in churning is even more to be avoided, because it is directly responsible for the following undesirable results.

1. Loss of butter fat in the buttermilk. When the churning temperature is high enough to reduce the churning period to about ten minutes, the loss of butter fat in the buttermilk may be as great as 1 or 2 per cent, whereas, under proper conditions, the loss usually does not exceed 0.2 per cent.

2. Injury to the quality of the butter.

(a) Too much buttermilk in the butter. When the butter granules are so soft that they do not remain distinct, but stick together in large masses, the washing out of the buttermilk is greatly interferred with and abnormally large quantities of it are incorporated into the butter. Butter of that kind has poor keeping qualities and quickly develops bad flavors. Other things being equal, the less buttermilk or curd in butter the better are its keeping qualities. The "off flavors" that so quickly develop in much of the farm-made butter are not produced by decomposition of the butter fat, but by decomposition of the milk solids which are found in the buttermilk. Because the drops of moisture pressed from the butter are milky in appearance, the butter is said to have a milky brine and for that reason is discriminated against.

(b) "Leaky" butter and too much moisture. Butter that "comes soft" retains large quantities of moisture from the buttermilk and wash water. Because of the softness of the butter the moisture is not well incorporated, but is found in pockets and large drops. Upon the butter standing some of the moisture oozes out, or, when the butter is cut, large drops appear on the cut surface. Such butter is said to be "leaky." That fault is objectionable in itself and has the additional objection of causing a material shrinkage in the weight of the butter.

(c) A weak, salvy body. Butter properly made has a firm, waxy body, but high temperatures during manufacture make it soft and of a salvy consistency. When eat-

(Continued on Page 103)

Feeding and Management of the Jersey Cow

By J. R. BOND, Manager.

INTRODUCTION To make an attempt to establish a set of rules governing the care and management of the Jersey cow, would be folly, and it isn't the purpose of this article to include anyone's particular condition, nor either is it wise to base these ideas as facts which will prove successful in various instances in which the ideas exhibited may be tried, but merely stating one's own idea of the various phases of management which has been tried and proved practical from the writer's standpoint of view. In this discussion the writer wishes to take the Jersey cow and point out the practical ideas of developing her, both from the standpoint of the breeder and the farmer.

I want to emphasize here at this time, the phase of personal interest which to my knowledge is one of the main factors which make the breeding of cattle profitable. There is a personal element in other words that enter into the care, feeding and management that determines the results that are obtained, and, if this element can not be applied properly, I fear that the results will be rather discouraging to one who tries and blame the cow, and, later, shoulders the results which makes her a total failure in the animal kingdom. It is a very common idea of the past that a cow is a machine in which to place different feeds, and in return receive the product most valuable to the human race. It is no doubt then that the element which is applied in so many different cases is the factor responsible for the wide variations of methods followed by so many different breeders.

CLASSES OF FOODS Animal foods may be devided into two classes, roughage and concentrates. Roughage includes all the coarse portions of the ration and concentrates embrace all the grain and mill products. From these two classes of foods we get the food of maintenance and that needed for production. Now to obtain the right proportions of food for maintenance and production it is necessary to formulate and balance all the rations used. The object in this is to provide sufficient bulk to satisfy the appetite, and, at the same time, furnish the proper amount of each ingredient needed for the individual and the work it is doing. If the ration lacks in bulk, the animal will be discontented, and if it contains nutrient to an excess needed for maintenance and work, there will be a gradual increase in flesh, but if it be lacking in either of the two constituents, there will be a gradual decrease in the amount of work, and more often a shrinkage in body weight which is very undesirable.

Rations may be classified under four different heads, and each one is as valuable as the other under conditions which they are adapted. These may be used very successfully at different stages of the development and their results are very distinct when administered properly. The following definition is given below to aid in getting the right idea as to the properties of each:

1. Nitrogenous or Narrow Ration—One that contains a high percentage of protein as compared with carbohydrates and fats.

2. Carbonaceous or Wide Ration—One that contains a high percentage of carbohydrates and fats and a very low percent of protein.

3. Balanced Ration—A feed or combination of feeds that furnish the several nutrients as carbohydrates, fats and protein in such proportions and amounts as will sufficiently and without excess of any nutrient, nourish an animal twenty-four hours.

4. Maintenance Ration—One that furnishes enough but no more, of each and all of the several nutrients than is required to maintain a given animal at rest, so that it will neither gain or lose in weight.

The term nutrient as used in the above definitions may be termed as that food constituent, that will aid in the support of life. These, however, must contain the same chemical composition and may be used in groups. From this we understand that there is a standard by which we are able to govern the amounts of nutrient and ratio of same, and from this we have what is termed, Nutritive Ratio, which may be applied in the methods of formulating rations for the animal to suit its needs in the different stages of development. In shorter words a feeding standard. By this we mean the ratio or proportion of protein to the combined carbohydrates and fats expressed in terms of Carbohydrates. Since one pound of fat will give 2.25 times as much heat and energy as one pound of carbohydrates, we determine the result by multiplying the fat by 2.25.

Below is the formula for determining the nutritive ratio of a feed:

Pounds Carbohydrates plus (pounds fat x2.25) pounds protein.

There are five principles to keep in mind in formulating rations and are very essential in the results that should be obtained from the combination. First, it must be palatable and such that will please the appetite of the cow at all times. One of the best methods of accomplishing this is to use the purest and best quality of grains in the mixture, also a sufficient number to furnish the desired variety. Second, the ration must be adapted to the animal, and in this case the Jersey. This includes the different phases of the animal over a year's time, and the feed that she needs to sustain her over the different changes that are made, such as the dry period, partuation, etc. Third, it must be adapted to the amount of work she is doing. This takes in the lactation period and more especially the cow that is on test. Fourth, feeds should be used that are prepared well before mixing, so that the animal may save time in the digestive operation, for if the feed is hard to masticate, it takes a longer time to render the nutrients in shape for the blood, and at the same time lessens the chances for thorough assimilation. Fifth, and one of the most important, it must be economical from the standpoint of the feeding. This I suspect has more to do with the production than either one of the other four, for the average man does not feed to the end of profitable production. If he does and the feed has cost him a good price per ton, he does not realize the profit from his feeding operations that every man wishes and therefore a general tendency to lessen the amount of feed to cut expenses.

The average man is continually asking himself, What can I feed my cow to make more profit? That problem is one worth considering, and every man is justified in taking plenty of time to figure out his rations to suit the conditions for which they are to be used. The results from this operation not only depend upon the feed, but the feeder and the individual consumer. There are three factors which determine to a great extent the success of the breeder, and these are the feed, feeder and the animal. It may be said that it pays to feed a good cow to the end of profitable production, but it never pays to feed a poor cow no matter how low the price of feed may be. This takes in the factor of culling and selecting and the maintenance of a profitable producing herd. The only accurate method of getting about this is through the methods of testing, selecting the profitable from the unprofitable and rendering yourself from ideas that are not thoroughly practical in the business. After this has been taken care of the remainder is left to the feed and the feeder.

The quantity of milk produced by the individual is dependent indirectly upon the constitution of the individual as fixed by breed and selection and directly upon feed, care and environment. The modern cow in the state of nature provided only for her offspring, but the same cow when years of selecting and careful breeding has been followed and proper amounts of feed given her responds quickly with far more milk than the calf will utilize. So generous is the good cow that few feeders feed to the end of profitable production and the quantity therefore is due to the feed and environment.

A knowledge of the proper amount of concentrates or grain which should be given to a cow is of great economic importance, and the lack of this knowledge is where so many stumble and lessen their chances for profit. Not only do or have they failed to study this problem, but they have lost sight of the specific purpose for which they are breeding. If a man is into the game for the commercial side of it there is nothing much desired but the quantity. If on the other hand the game is a breeding operation, there are several factors which determine his success. He must first be well informed on the habits and characteristics of the family he is breeding, and then arrange his methods so as to receive the proper results from the operations. It is not necessary at this time to go into detail concerning these operations which are factors in the developing of the herd and maintaining same for they will be discussed under the following heads, giving in detail proper methods of handling the herd.

THE JERSEY CALF One of the important factors in raising calves is getting results from proper methods which should be used in securing the correct kind of type, growth, etc. It is often said that type will take care of itself if proper mating is made between the sire and the dam, but I am of the opinion that there is more type made than bred, and the calf will in most cases develop according to its feed and environment. The one idea that is predominate is to make it a better individual than either the sire or the dam. This can only be done by proper handling, stimulated by the correct kind of feeding and proper environment.

At the time of partuation the calf should be nourished immediately by letting it have access to the colostrum milk from the udder of the mother. This serves two purposes, it relieves the mother of some of the inflamation that might be in the udder and has a stimulating effect upon the calf.

(Continued on Page 110)

(Continued from Page 69)

the two following methods, should be attempted only by those who are properly qualified to do the work.

THE INTRADERMAL TEST (INTO THE SKIN) The intradermal test for detecting tuberculosis is used to a considerable extent, especially on range cattle not easily controlled. When made by those who have become skilled in its application, it is very accurate. In this test the tuberculin is injected between the layers of the skin, only a few drops being used, and it is usually applied in the region at the base of the tail, where the skin is soft and nearly hairless. The intradermal test is satisfactory also for the diagnosis of tuberculosis in swine and, when so used, the tuberculin is applied into the skin of the ear near its base.

The reaction from the intradermal test consists of a swelling at the point of injection and is observed from 48 to 108 hours after the injection. The character of the swelling varies, and a proper diagnosis of tuberculosis by this test can be made only by an experienced person.

THE OPHTHALMIC TEST (INTO THE EYE) Still another method, known as the ophthalmic, is used to some extent and has been found to be of considerable value in what is known as "check" testing; that is, it is used in connection with either of the previously described methods. Sometimes a tuberculous animal that fails to react to those tests shows evidence of the disease upon the application of the ophthalmic test. The ophthalmic tuberculin is placed in one eye and the other eye is used as a check. A reaction is indicated by a characteristic discharge from the eye receiving the treatment, which may occur in from three to ten hours after the application or even later. Some swelling and inflammation of the eye and lids are often noted.

RELIABILITY OF THE PRINCIPAL TUBERCULIN TESTS The subcutaneous test is the principal test used by the U. S. Bureau of Animal Industry. It consists in injecting, under the skin, a small quantity of tuberculin. Carefully and conscientiously applied, with good judgment exercised in both its administration and interpretation, it is wholly effective.

The intradermal test is recognized by the bureau on strictly range cattle or animals whose movements are difficult to control, and in area work. In this test the tuberculin is injected between the layers of the skin.

The ophthalmic test, or so-called "eye test," is not at present accepted for testing cattle for interstate shipment, though it has value as a check test and is recognized for that purpose. It is applied by placing the ophthalmic tuberculin in one eye, using the other as a check. The ophthalmic test has given best results under farm conditions or in other cases where the eyes are normal. For testing cattle in transit or in the stockyard the test is less dependable, owing to the fact that the eyes may be abnormal as the result of irritation or injury from dust, cinders, or other results of transit. In all cases the tests, used either alone or in combination, should be applied by capable persons familiar with tuberculin testing.

POST-MORTEM APPEARANCES Animals affected with tuberculosis may show the effects of the disease in almost any part of the body. In advanced cases the lesions are easily found, but when the disease is of recent origin or if a slightly diseased area has been encapsulated or closed up, it is often very difficult to find evidence of the disease. Lesions in advanced cases generally appear as nodules or lumps, which are tubercles formed as a result of the disease. These lumps may be found in great numbers in the lungs and abdominal organs. The lesions are of various sizes and may contain pus, either soft or hard; many times it is gritty. Tubercles are often found in various numbers attached to the walls of the thoracic and abdominal cavities. Lesions of the disease also occur in the lungs, liver, and spleen. The lymph glands, to some extent, are usually affected, and, when cut into, show diseased areas characteristic of tuberculosis.

Lesions of the disease may be found also in the skin and in or on the bones. In animals only slightly diseased, the lesion may be hidden so that it is impossible for even the most skillful person in post-mortem work to find it. A microscopic examination of the lymphatic glands or other tissues often reveals the presence of tubercle bacilli when no lesions can be seen by the naked eye, a condition showing that the disease is just starting. When animals have reacted to the tuberculin test, a very careful post-mortem examination should be made. The action of tuberculin is often discredited when on post-mortem the lesions are not plainly seen, but experience of many years has shown that very few animals reacting to the test were not affected with tuberculosis to some extent even though some were very slight.

(Continued on Page 104)

MY PERSONAL EXPERIENCE
(Continued from Page 27)

1879 we were fortunate in selling fifty acres of the farm, including all the buildings, for a public institution, for $5,000. This put us out of debt, but no home, so we rented, the spring of 1880, a 250-acre farm, cash rent—$400 per year for five years. About one-third timber, land poor and poorer fences and buildings. No stabling for cows.

With two yoke of oxen and a team of horses we went to the timber, cut logs and hauled to a nearby sawmill for lumber for a cow stable, which we built by planting four rows of poles ten feet apart, spiking on 2x4 scantling for stringers, boarded up and down for outside weatherboarding, put in rigid stanchions, a platform with drop for the cows to stand on, a loft overhead for hay and grain, a clapboard roof rived from the best splitting oak we could find. This stable was 28x50 feet square, holding 28 cows.

There was a good spring and a partly fallen down springhouse near the house which we patched up the best we could. We needed ice for cooling milk, so went to the saw mill, got slabs with which we made an ice house 12x14 feet square, packing the ice, leaving about one foot around the outside for sawdust. Giving this careful attention we kept ice all summer. This completed our buildings on a rented farm, all made at our expense. Since then we have had much better buildings and equipment, but never have we attained greater success than we did on that rented farm. What we lacked in buildings and equipment we tried to make up by close attention to our work.

During the fall of 1881 we showed some cattle at the Wheeling (W. Va.) State Fair. There I met Mr. Axtell, editor and founder of the National Stockman and Farmer of Pittsburgh. Mr. Axtell noticed my intense interest in my cattle, became interested in me. Some months later Mr. Axtel wrote me, asking that I send to Spahr & Milligan, East End grocers, a sample shipment of our butter. We first shipped in two pound fiber boxes, nothing said about price, but returns came every month and very much better than we were getting in our local market.

We soon sent to the Vermont Farm Machine Co. for a square block printer and shipping boxes with trays and a metal box in center for ice. This greatly pleased the grocers. This so far as I am able to learn, was the first square block print butter that ever went into the Pittsburgh market.

This grocery firm finally offered us 35 cents per pound for all the butter we could make, the year round. This offer we fully appreciated, put renewed vigor into all the Baileys; all did our very best and held that trade for many years—as long as we made butter.

Remaining on this farm for six years we rented another farm, on which we spent about $150 in arranging stables. No spring house, so we made dugout or cave creamery on hillside, siphoning water from a well above. We put in a butter worker and barrel churn.

Butter made in this little dugout shown at the first National Dairy Show ever held in the United States, 1887, at Madison Square Garden, New York, won second prize, competing with over 200 entries of private dairy butter made by the fine butter makers of the East. This success caused a love feast in the Bailey home home, added ten years more to my life, restored the bloom of youth to my wife's cheeks and made the little Baileys leap for joy.

The spring of 1888 we bought the last rented farm, have it now pretty well equipped and fully stocked with pure-bred Jersey cattle.

We are both very much interested in our Jerseys, and our farm and our community, yet we feel that by far our greatest achievement is in raising a family of four boys and two girls, all happily married—22 grandchildren and two great grandchildren, and, better yet, all so far interested in some branch of the dairy business.

Celebrated the 50th anniversary of our marriage July 26, 1921.

MAKING BUTTER ON THE FARM
(Continued from Page 99)

en it seems to melt slowly and stick to the mouth, in contrast to the quickly melting and quickly disappearing butter with a firm waxy body.

The use of the proper churning temperature is therefore essential to the production of first-class butter, which means that the churning period must occupy thirty or forty minutes. There is no short cut in churning. Patent churns that churn butter in seven minutes produce practically the same harmful results as those just described.

When cream is ready for churning the churn should be prepared. It should be cleaned thoroughly, rinsed with scalding water, then thoroughly rinses and chilled with cold water.

The butter ladles, paddles, worker and printer should be treated in the same way, and all but the worker placed in a pail of cold water until needed. If that is not done, the butter will stick to them.

PUTTING THE CREAM INTO THE CHURN Cream should be poured into the churn through a strainer to break up possible lumps and to re-

(Continued on Page 107)

THE OHIO JERSEY  NINETEEN TWENTY-TWO

(Continued from Page 102)

METHODS OF ERADICATION Cattle owners who do not know whether tuberculosis exists among their animals should ascertain the fact by having them tuberculin-tested and physically examined by a qualified veterinarian. In many cases thousands of dollars and very valuable breeding animals might have been saved by taking up tuberculosis-eradication work in time. Three main projects comprise the general campaign of eradication, as follows:

1. Eradication of tuberculosis from pure-bred herds of cattle.

2. Eradication of tuberculosis from circumscribed areas.

3. Eradication of tuberculosis from swine.

It is important to eradicate tuberculosis from pure-bred herds of cattle at the earliest possible date because the spread of the disease is greater among such animals than among grade cattle. The reason is plain; pure-bred animals are shipped extensively to every part of the United States for breeding purposes. A pure-bred bull or cow may be shipped from Maine to Texas, or from the State of Washington to Florida. If it is diseased and is introduced into a healthy herd, it not only fails to fulfil the purpose for which it was intended—the upbreeding of the herd—but it actually causes heavy damage by spreading the disease to healthy animals.

ACCREDITED-HERD OR HONOR-ROLL PLAN The breeders of pure-bred registered cattle fully appreciated the above-mentioned fact when, together with the livestock sanitary officials of practically all the States, they adopted what is known as the accredited-herd plan, the principals of which are that herds found to be free from tuberculosis on two successive annual tests are placed on the Honor Roll, and a certificate is given to the owner by the State and the Federal Government. The certificate entitles animals of that herd to be shipped interstate without further tuberculin testing for a period of one year. This plan is becoming well known to breeders throughout the United States.

The methods of eradicating tuberculosis from grade herds are, of course, the same as for pure breds. No owner can rest assured that his herd is free from tuberculosis unless it has been properly tuberculin tested. To make a satisfactory test all the cattle should be in normal condition and, so far as practical, the cattle should be stabled under usual conditions and among usual surroundings. Feeding and watering should be conducted in the customary manner, with the exception that feed and water should be given only after the temperature has been taken. Careful physical examination of each animal should be made before or during the application of the test. If animals react to the test they must be separated from the rest of the herd.

PRINCIPAL BENEFITS OF COMPLETE ERADICATION OF TUBERCULOSIS Increased value of individual animal and increased herd value. Ability to ship interstate from accredited herds without further testing for a period of one year.

The owner's name being listed in pamphlets published by the respective States and the Bureau of Animal Industry.

Confidence of those who desire to purchase cattle to add to their herds.

Satisfaction of knowing that the dairy products offered for sale are free from diseased germs.

Elimination of economic losses caused by the disease.

ERADICATION FROM AREAS The unit territory to be worked will depend mainly upon the extent to which the disease has spread. In some States practically every county contains numerous tuberculosis herds, so that, to control the disease effectively, all the herds must be tuberculin tested. In other States, however, the disease is confined to the beef and dairy herds recently established or to which new animals have lately been added. In the latter case it would not be necessary to test all the cattle, but the examination of the herds should be sufficient to demonstrate most satisfactorily that no diseased herds are overlooked. This can be done by testing several herds in each section of the country wherever there is a suspicion that the disease may exist.

As a general plan, it is best to take up the work by counties, and substantial cooperation should be obtained from the county government. Each county should pay (1) part of the expense of exterminating the disease by employing inspectors to make the tests, (2) part of the indemnities paid for tuberculous animals, and (3) its share of the cost of cleaning and disinfecting infected barns, stables, and sheds. When a large percentage of the herds of a county are diseased, it is advisable to clean up the herds within a township or possibly one-third or one-half the area. The progress depends upon the degree of infection found and the cooperation furnished by the owners.

In 1910 the Bureau of Animal Industry took up the eradication of tuberculosis from the herds in the District of Columbia, which

COME TO OHIO FOR GOOD JERSEYS — THREE THOUSAND BREEDERS WAITING TO SERVE YOU.

has an area of 60 square miles. At that time, 1,701 cattle were found. Every animal was tuberculin tested; of the total number, 321 cattle or 18.87 per cent, were tuberculous. The reactors were removed from the herds and, in most instances, were slaughtered. The infected barns, sheds, and premises were cleaned and disinfected.

Each year since the inauguration of the campaign all the cattle have been tuberculin tested with the result that the infection has almost disappeared.

MEASURES OF PREVENTION Since, after many years of study and experience, no satisfactory cure for tuberculosis among animals has been found, prevention of the disease is extremely important. State and Federal Governments have made vigorous efforts to stop the spread of the disease by regulating the movements of cattle, and recently, with that object in view, action has been taken in some localities to regulate the movement of cattle from one county to another. Regulation of intercounty movement should be encouraged because it brings the matter nearer home to the live-stock owner. It is he who must take a very important part all through the campaign of tuberculosis eradication, and if he is in favor of measures to prevent the spread of the disease and faithfully abides by those measures, eradication will be acomplished more speedily.

From what has been said already about the dangers of shipping diseased cattle, it is plain that the movement of tuberculous cattle, except for immediate slaughter or to quarantine, must be stopped whenever possible. After diseased animals are found and removed from the premises, a very thorough cleaning and washing of the inside of the barn and other buildings where the animals have been should be made. This must be followed by the proper application of some approved disinfectant. The use of disinfectants without first doing the necessary and proper cleaning is ineffective for the reason that the germs of the disease must be exposed. All utensils or anything else that may have become contaminated by use around the diseased animals should likewise be cleaned and disinfected. The manure and refuse must be hauled from barnyards or lots to plowed fields, spread thin, and exposed to the sunlight. The yards and lots, including feed troughs, water troughs, and fences can then be sprayed properly with the disinfectant.

All this means much work but it must be done to prevent infection from spreading to the healthy animals. Proper sanitary conditions on premises where live stock are kept is of great importance in keeping the animals healthy and able to resist disease. Sanitation, in its broad sense, includes the admission of abundant sunlight and fresh air properly regulated.

DISPOSAL OF REACTORS The disposal of reactors depends largely upon the State laws and live-stock regulations of the State in which the herd belongs. If the animals are pure bred and registered and of unusual merit, they should be segregated, preferably on farms set aside by the State or by the pure-bred cattle associations, for the purpose of retaining tuberculous cattle in quarantine. If the condemned animals are grade cattle, or pure breds not especially valuable for breeding, it will probably be more economical to have them slaughtered than to hold them in quarantine. Of course the fact is recognized that in States and communities where tuberculosis exists extensively the slaughtering of all reactors is impracticable. In such instances the infection can be reduced on all the farms by keeping the tuberculous animals separate from the healthy ones. The tuberculous cattle are kept under quarantine restrictions until no longer profitable; meanwhile the healthy animals are safe from the danger of infection.

RETESTING It is rather uncommon for tuberculosis to be eradicated from an infected herd after one tuberculin test. After the removal of reactors the herd should be retested at the expiration of six months, and a second retest may be advisable six months later, but the practice of testing herds more frequently than that is not usually advised. After two or three semi-annual tests the herd should not be tested oftener than every twelve months. While the subcutaneous test is considered preferable for gentle cattle, the ophthalmic and intradermal methods of testing may be employed to advantage as adjuncts to it; and it is believed that in some instances herds may be freed of tuberculosis earlier by the judicious use of all three methods. No general outline can be given as to when all three tests should be employed, the matter should be left to the judgment of the veterinarian under whose direction the work is carried on.

THE ACCREDITED- HERD PLAN— EVERY CATTLE OWNER ELIGIBLE A tuberculosis-free accredited herd is one which has been tuberculin tested by the subcutaneous method or any other approved method under the supervision of the Bureau of Animal Industry, or of a regularly employed veterinary inspector of the State in which co-operative tuberculosis eradication is being conducted.

It is a herd in which no animal affected with tuberculosis has been found upon two

(Continued on Page 106)

annual or three semi-annual tests, made as described, and by physical examination.

Owners of tuberculosis-free herds receive a certificate issued by the bureau of the State live-stock sanitary authorities. The certificate is good for one year from the date of the test, unless revoked at an earlier date.

MARKING ANIMALS FOR IDENTIFICATION It is very important to mark properly all cattle which react to the tuberculin test, so that they may be easily identified as affected with tuberculosis. One method that is quite generally used is that of branding; a letter "T" about 2 inches high is branded on the lower jaw, or sometimes it is placed on other parts of the body where it can be seen readily. In addition to the branding it is advisable to tag the reacting animals so that one may be identified from another, and in that way the results of the post-mortem examination can be connected up with the reporting of the tuberculin tests. The tag is usually placed in the ear of the animal and contains a serial number as well as the word "Reactor."

Another method that is sometimes used is the punching of a letter "T" out of the ear, and it has been quite satisfactory.

The marking of cattle that have passed the tuberculin test is a matter that is being handled in different ways throughout the United States, and it is believed that the present methods of marking will be improved. In some cases a metal ear tag is used and in others certain marks of identification are tattooed in the ear. Tattooing has an advantage over the tagging in that it is less expensive and probably more permanent.

It is not often necessary to require special marks on pure-bred registered cattle, as the owner usually has a method of identification, and this method of marking can be used in connection with the tuberculin test as a record; but on grade animals it is desirable to use some system of marking that will show that the cattle have been tested and found apparently free from tuberculosis.

A system of marking swine so that the origin of those found to be tuberculous on post-mortem examination can be learned would be very desirable, and some experimental work along this line is now being conducted by the bureau in co-operation with owners and packers. The results so far obtained indicate that the tattooing of a number or some mark of identification into the skin of the hog is the most practicable method, but of course it entails additional labor and expense which would amount to a great deal if carried on throughout the country. It is hoped a more practicable and economical means of marking swine for identification will be developed in the future, as much good in the campaign for the eradication of tuberculosis in live stock can be accomplished in that way.

Many shipments which contain tuberculous swine are traced back to the farm by a system of reports kept by the Bureau of Animal Industry. By developing the methods of tracing tuberculous cattle and swine from the abattoir back to the farm where they were raised, efforts can be directed in eradicating the disease from these herds.

APPRAISEMENT AND INDEMNITY In addition to the various benefits derived from eradicating tuberculosis, provision for the appraisement of diseased cattle with indemnity for those slaughtered is a further incentive. Federal legislation and supplementary laws in numerous States now divide the burden of loss, so that the Government, the State, and the owner of the cattle all bear a share of it. Briefly, the Federal law provides that the Secretary of Agriculture may reimburse partly owners of animals destroyed on account of tuberculosis in co-operation with States, counties, and municipalities. The bases upon which Federal indemnities are paid are:

1. No payment shall exceed one-third of the difference between the appraised value of such animal and the salvage value.
2. No payment shall exceed the amount paid or to be paid by the State, county, or municipality.
3. In no case shall any payment be more than $25 for any grade animal or more than $50 for any pure-bred animal.
4. No payment shall be made unless the owner has complied with all lawful quarantine regulations.

WHY THE TUBERCULIN TEST IS HARMLESS Tuberculin contains only the sterilized products of tubercle bacilli. It does not contain any living germ; therefore it is harmless to any animal whether healthy or diseased.

HOW THE WORK OF ERADICATION IS DIVIDED BETWEEN STATE AND GOVERNMENT OFFICIALS AND OTHERS Tuberculosis eradication is a co-operative work of the Bureau of Animal Industry of the U. S. Department of Agriculture, the live stock sanitary officials of the various States, and individual cattle owners. The bureau and State officials send veterinary inspectors to apply tuberculin tests to the herds of those owners who sign a co-operative agreement, which places their herds jointly under the supervision for the control and eradication of the disease.

BUTTER MAKING ON THE FARM
(Continued from Page 103)

mope curd particles and any foreign matter that may be in it. In order to have the necessary concussion the churn should be only about one-third full. If too full, the churning period is prolonged, and if the cream foams it nearly fills the churn and prevents concussion. In that case it is usually necessary to remove some of the cream in order to obtain butter in a reasonable time.

ADDING BUTTER COLOR Except late in the spring and early in the summer, when butter has a naturally high color, a small quantity of butter color is usually added. In winter the quantity required to produce a shade of yellow like the desirable June color varies from about twenty to thirty-five drops per gallon of cream.

The color having been added to the cream, the churn may be started at a speed to produce the greatest concussion, which can be determined largely by the sound. About sixty revolutions a minute is the usual speed for the common barrel type of churn. After a few revolutions the churn should be stopped, bottom up, and the cork removed to permit the escape of gas. This is repeated two or three times in the early stages of churning. At that period cream produces a very liquid sound, and the glass in the churn is evenly covered with cream.

When churning is nearly completed there is a noticeable difference in the sound made by the cream, while on the glass in the churn a thick, mushy mass will appear, which occasionally breaks away, leaving the glass clear. At this point the butter granules are just forming, and the cream is thick and finely granular, like yellow cornmeal mush, with buttermilk separating slightly from the tiny granules. The churn should be revolved several times, then stopped and the butter examined in order to prevent over-churning. When the granules are the size of grains of wheat the churning is completed. To continue the churning until the butter is in large masses is a bad practice, because it incorporates quantities of buttermilk which can not be washed out. The bad effect of too much buttermilk in the butter has been discussed already.

Churning completed, the buttermilk is drawn through the hole at the bottom of the churn and is run through a strainer to catch any particles of butter.

WASHING, SALTING AND WORKING THE BUTTER The object of washing butter is to remove the buttermilk. The only way that this can be done properly is to wash the butter when it is in small granules so that the largest possible surface is exposed to the water. To try to remove buttermilk by working it out of the butter is not effective; moreover, the excessive working injures the grain and body of the butter.

While the last of the buttermilk is draining off the wash water should be prepared. Only pure, clean wash water should be used, and it should be twice the quantity of and at about the same temperature as the buttermilk. The water should be placed in a pail or other receptacle and its temperature determined with a thermometer; if necessary it should be tempered by the addition of either warm water or ice. If the butter granules are too soft or too hard the temperature of the wash water may be either a buttermilk. Warm water has the same effew degrees warmer or colder than the fect upon the body of the butter as high churning temperatures, whereas cold water makes the butter so hard that it can be worked only with great difficulty, and if very cold the proper incorporation of the salt is practically impossible.

After the buttermilk has been drawn off, the cork is replaced and one-half the wash water is poured into the churn. The cover of the churn is then replaced and the churn given about four rapid revolutions. The wash water is drawn off and the washing repeated. Two washings are usually sufficient, the second wash water when drawn off usually being almost perfectly clear.

SALTING AND WORKING While the wash water is draining off the worker should be rinsed again with hot water followed by a thorough rinsing and cooling with cool water. This must be done immediately before using, because if the worker is slightly dry the butter will stick to it. The lever worker is widely used and gives satisfactory results, though other types do just as good work.

The butter, which is still in the granular condition, is removed from the churn with the ladle and placed in a convenient receptacle for weighing. The old-fashioned butter bowl is convenient, and this is the only use that should be made of it. The butter having been weighed the quantity of salt is calculated on the basis of three-fourths of an ounce for each pound of butter. The quantity may be varied to suit personal taste or the requirement of the market. The best grade of butter salt or table salt should be used. The butter is placed upon the worker, spread out about 2 inches thick, and the salt, free of lumps, sifted upon it. The butter is then pressed with the lever or other device, care being taken to press and not to rub or smear it. After being pressed into a thin layer it is folded upon itself into a pile and the press-

(Continued on Page 108)

ing repeated. The working is continued until there is a thorough and even distribution of the salt and a desirable grain and body have been produced.

The working of the butter is a very important step in the making process and should receive careful attention. Too much working is a common fault in farm-made butter. Overworked butter has a sticky and salvy body, a dull, greasy appearance and gummy grain. It feels warm in the mouth, sticks and dissolves slowly. Properly worked butter has a waxy body and a bright appearance, and feels cool and dissolves quickly in the mouth. Butter has a proper grain if a slab breaks when bent at an angle of about 45 degrees and the broken surface has the appearance of broken steel. In addition, overworking butter injures its keeping properties.

When butter is underworked it is brittle, may be gritty because of undissolved salt, and, worst of all, may be mottled or uneven in color. The latter fault is common, and the purchaser who is not versed in butter making sometimes thinks it is due to mixing light and deep-colored butter, and for that reason mottled butter is strongly discriminated against on the market. Mottles are caused by the uneven distribution of salt, the deeper-colored streaks or spots containing more salt than those of lighter color. To prevent that condition the butter must be worked until the salt is evenly distributed throughout the butter. Butter that is cold and very firm requires more working than when it is comparatively soft.

BUTTER PACKAGES For home use butter is frequently packed into glazed earthenware crocks, which are very satisfactory and convenient receptacles for butter on the farm. If the glazing is imperfect, however, the crocks absorb butter and soon become very unsanitary.

For market the rectangular 1-lb. print is the most desirable form. It presents a more attractive appearance than the crock or "country roll," is more convenient and easily handled, and can be inserted into a carton which not only protects the butter but also adds greatly to the appearance of the package. To make prints, the printer is pressed upon the butter on the table until it is completely filled, the surplus is then scraped off with the paddle and the print pressed out on parchment wrapping paper. In order that the prints may weigh exactly a pound the printer should be carefully regulated and an occasional print weighed on an accurate scale. Prints for market should be wrapped in white parchment paper made for the purpose, 8 by 11 inches in size, and placed in paraffined cartons, upon which may appear the name of the farm or brand.

After printing and wrapping, the butter should be placed in a refrigerator or other cool place.

The churning utensils should then receive immediate attention. They should all be thoroughly cleaned by means of a hand brush, hot water and dairy cleanser or washing powder, and thoroughly rinsed with boiling water. To place them in the sunshine or occasionally wash them with limewater aids in keeping them sweet.

DIFFICULT CHURNING The farm buttermaker sometimes fails to obtain butter after churning the usual length of time; in fact, the churning is sometimes prolonged for several hours without obtaining butter. The causes of the difficulty, together with the remedies, are as follows:

1. Churning temperature too low. It may be necessary, under exceptional conditions, to raise it to between 65 degrees and 70 degrees Fahrenheit.
2. Cream too thin or too rich. It should contain about 30 per cent butter fat.
3. Cream too sweet. If ripened to a moderate acidity it will churn more easily.
4. Churn too full. In order to obtain the maximum concussion the churn should be not more than one-third full.
5. Ropy fermentation of the cream preventing concussion. This may be prevented by sterilizing all the utensils and producing the milk and cream under the most sanitary conditions. If additional measures are needed, the pasteurization of the cream, with subsequent protection from contamination, and ripening it with a good starter, will be effective.
6. Individuality of the cow. The only remedy is to obtain cream from a cow recently fresh, or cream that is known to churn easily, and before ripening mix it with the cream that is difficult to churn.
7. The cow being far advanced in the period of lactation. The effects may be at least partially overcome by adding, before ripening, some cream from a cow that is not far advanced in the period of lactation.
8. Feeds that produce hard fat. Such feeds are cottonseed meal and timothy hay. Linseed meal, gluten feed and succulent feeds such as silage and roots tend to overcome the condition.

EQUIPMENT FOR FARM BUTTERMAKING The following equipment is needed for buttermaking on the farm:
1. Milk pails.—They should be of the type commonly known as covered-top, should be heavily tinned, and have all seams flushed with solder so that they can be cleaned easily.
2. Cream separator.—Any make is satisfactory if it skims clean and can be thor-

oughly cleaned and sterilized.

3. Shotgun cans.—As a cream container the style of can known as the "shotgun can" is much to be preferred to crocks and many other types of cans and pails commonly used. This can usually measures about 8½ inches in diameter and 20 inches high. These cans are easily handled, covered and cleaned.

4. Cream-cooling tank.—Where there is an abundance of cold water, any tank, properly used, will be effective. In very warm climates or where cold water can not be run through the tank several times daily, or where ice is used, it is advisable to use an insulated tank.

5. Churn.—The barrel type of churn is simple, inexpensive, easy to operate and easily cleaned.

6. Butterworker.—The lever and the table butterworkers are very satisfactory. The former is simpler and less expensive. When making large quantities of butter a table worker or combined churn and worker is frequently used.

7. Thermometer.—A floating dairy thermometer should be used.

8. Cream and buttermilk strainer.—A strainer similar to a colander or a strainer dipper is frequently used for straining both the cream and buttermilk. A hair sieve is sometimes used as a buttermilk strainer because butter does not stick to it as it does to tinware.

9. Cream-stirring rod.—A rod with a 4- or 5-inch disk on one end is more effective in stirring cream than a spoon or other implement. Stirring rods should be well tinned and smooth so that they may be cleaned easily.

10. Wooden paddle.
11. Wooden ladle.
12. Tin pails.
13. Half-gallon tin dipper.
14. Hand butter printer.
15. Scrub brush.—A stiff fiber brush is preferable to one with soft bristles.

PLAN OF DAIRY HOUSE A conveniently arranged dairy house is very desirable in making butter on the farm, especially if any considerable quantity is produced. A small, simple building will usually answer the purpose.

SUMMARY OF STEPS IN MAKING BUTTER ON THE FARM
1. Produce clean milk and cream. Cool the cream immediately after it comes from the separator. Clean and sterilize all utensils.

2. Ripen the cream at from 65 degrees to 75 degrees Fahrenheit until mildly sour. Always use a thermometer in order to know that the right temperature is reached.

3. Cool the cream to churning temperature or below, and hold at that temperature for at least two hours before churning.

4. Use a churning temperature—usually between 52 degrees and 66 degrees Fahrenheit — that will require 30 or 40 minutes to obtain butter.

5. Clean and scald the churn, then half fill it with cold water and revolve until churn is thoroughly cooled, after which empty the water.

6. Pour the cream into the churn through a strainer.

7. Add butter color—from 20 to 35 drops to a gallon of cream—except late in the spring and early in the summer.

8. Put the cover on tight; revolve the churn several times; stop with bottom up, and remove stopper to permit escape of gas; repeat until no more gas forms.

9. Continue churning until butter granules are formed the size of grains of wheat.

10. Draw off the buttermilk through the hole at the bottom of the churn, using a strainer to catch particles of butter. When the buttermilk has drained out, replace the cork.

11. Prepare twice as much wash water as there is buttermilk, and at about the same temperature. Use the thermometer; do not guess at temperatures. Put one-half the water into the churn with the butter.

12. Replace the cover and revolve the churn rapidly a few times, then draw off the water. Repeat the washing with the remainder of the water.

13. The butter should still be in granular form when the washing is completed.

14. Weigh the butter.

15. Place the butter on the worker and add salt at the rate of three-quarters of an ounce to a pound of butter.

16. Work the butter until the salt is dissolved and evenly distributed. Do not overwork.

17. Pack in any convenient form for home use, or make into 1-lb prints for market, wrapping the butter in white parchment paper and inclosing in a paraffined carton.

18. Clean the churn and all butter-making utensils.

COW-TESTING ASSOCIATIONS' RECORDS SHOW:

Butter-fat
The Mature Cows (Grade and pure bred of all breeds) in 120 Cow-Testing Associations averaged   264 lbs.
The **Pure Bred Jerseys** in these Associations averaged ....   **304 lbs.**
(Figures by U. S. Dept. of Agriculture)
The more Pure Bred Jersey blood in your herd the greater the production.

FEEDING AND MANAGEMENT OF THE JERSEY COW
(Continued from Page 101)

At this time I have found it advisable to watch the calf and prevent it from taking too much milk in order to keep down scours. This I handle by placing the calf in a small crate which can be placed in the corner of the maternity stall. That I think has two important results, one in giving the mother a little rest, because as the calf is moving around it causes more or less excitement, later developing into a state of nervousness on the part of the mother. Second, I have found it much better to keep the calf quiet, and this affords a valuable asset in the future strength and growth. For the first three days I would let the calf suck at regular times, depending altogether upon the condition of the dam, and often it is advisable to let it go longer where the is a great deal of inflamation in the udder of the mother. Under normal conditions I prefer to take the calf away at this time and feed on the bucket. I always prefer the milk from the mother, and regulate the quantity according to the vigor of the calf. Ordinarially I start the calf on six pounds of milk a day and gradually increase the amount until I have reached the maximum of its feeding capacity. In all cases, however, these operations are adjusted to the surrounding conditions, and personal judgment goes a long ways in the results obtained.

As the calf grows older I prefer a gradual change in its feed or milk, and this I make through adding skim milk in small amounts. At no time do I prefer to feed the calf more than eight pounds of whole milk a day, but get my quantity by adding skim milk until finally at the end of about six to eight weeks I have the calf on pure skim milk. At this time of the calf's life where fed on the bucket, is where the important phase of the future success of the calf comes in. It is at this time that you start your proper type, and you can either make it a scrub or through proper methods of feeding, make it a good one. During the time you are making the change in the milk it is wise to learn it to eat grain. This may be easily done by putting into the bucket where it has just finished its milk a little bran and oats, equal parts. The calf will soon learn to eat its grain then right after it has finished its milk and can later be changed to a pan or small box. In addition to the skim milk that the calf is getting it is a good plan to place in the milk about one teaspoonful of oil meal. This gives the calf a nice finish and a very healthful appearance.

GRAIN RATIONS FOR CALVES Equal parts of bran and oats with one part shelled corn in the amounts that it will eat and clean up well, with all the clean timothy hay in front of it all the time, will satisfy the calf until about three months of age. During this time the calf should be getting about fifteen to twenty pounds of skim milk. After this time I prefer individual feeding in proper amounts to keep it growing well. The remark has often been made that the first six months of a calf's life is the determining factor in its future value. That is true, and, if started correctly, will develop accordingly. In case the calf is a bull I prefer to force him as fast as possible to get that vigorous amount of growth that you want on the young bull. Scouring in calves is the most dreadful trouble that the calf feeder has to contend with and a factor that is easiest controlled by him if proper methods are followed. The first and best policy is to keep the calf in a well cleaned stall where plenty of sunlight can have access. This old saying that an ounce of prevention is worth a pound of cure, works pretty well here for once a calf gets sick it will trouble the feeder until it is several months of age. Good, substantial feeds, with plenty of exercise, given regularly, will prevent most of this trouble.

The care and management of the calf during the first few months of its life, determines to a great extent the value of the individual at serviceable age. The calf from six months old in case of the heifer should receive feed that will keep it growing well, and, at the same time, prevent it from accumulating fat for it is not wise to keep the dairy calf fat. At the time of breeding, and that usually should be from the ages of fourteen to sixteen months old, she should have plenty of size so that the remainder of the justation period can be used in developing the fetus. If the heifer has to grow during this time the chances are that her first lactation period will be unprofitable.

If in other words the heifer has been grown out properly, she will undoubtedly make a good showing with her first calf.

One of the important factors in maintaining a good herd is the uniformity of udder conformation, and this can only be obtained by proper methods of development and prevention in udder sucking. If the heifers are allowed to suck each other, chances are that with their first calves they will have very rugged udders and in several cases I have seen blind quarters as the result. Force feeding, often occuring where the young things are fitted for show, causes milk secretion and growth of the udder. Especially does this occur in the ages of senior yearlings where they have been bred. This should be prevented, and only can be by the proper method of feeding and with feeds of low protein content. The young bull should be pushed right along for the first few months for proper feeding and growth may show up something in his characteristics that is very desirable. Many bulls have been ruined from lack of care, while young, and these things occur in good herds as well as the weaker ones. Bulls are hard to sell, and the ability to sell the offspring from the herd is one of the determining factors of success in the breeding operation. One of the good assets in having something that the public wants is to have good appearance shown on the animals. Training the bull to handle well and showing the results of the proper feeding that has made him that kind of an individual is a drawing card to the buyer, not only this but that you are getting returns from the methods used. These two factors of training and development go hand in hand, and are dependable upon each other for proper results.

At the time of the first calf, if the heifer has had proper care and feed, she will have a nice udder, backed up by a good capacity to handle food to support her first lactation period. Great care should be exercised in handling the heifer at this phase of her life, for this being the turning point determines to a great extent her usefulness as a producer. Taking the cow at this time, the operations may well be devided into eight different classes, and a discussion of each will pretty well include the important factors in making a cow a success.

The eight factors to be considered are: First, fitting before the calving period; second, care immediately before partuation; third, care during partuation; fourth, care immediately after partuation; fifth, period between partuation and starting on lactation period; sixth, getting the cow on full ration; seventh, rations used; eighth, rebreeding.

In the process of fitting for the calving period a number of things are to be considered. One of the most important is that of drying up the cow. There is a wide variation of methods along this line, but I prefer to skip milkings, once every day until she stops the secretion of milk. This is a long drawn-out process, but with care it keeps down diseases of the udder, such as garget and mimmitis which are likely to destroy the quarter or quarters of the udder. I like to have the cow dry about six to eight weeks. During this time I take the protein feeds away from her and place her on what may be termed a dry ration. The mixture that I have found very good for this time is:

Two parts of ground oats.
Two parts of bran.
One part of oil meal.

In some instances I prefer to add one part of ground corn where the cow is low in flesh. Considering the cow in normal condition the above ration will suffice, if fed from six to eight pounds per day and more, depending on the size of the cow. All the roughage should be taken away at this time, except good alfalfa or clover hay, and she should have all this that she cares for. In about one week of partuation I perfer to change the ration from the above mixture to pure bran and oil meal, fed in the form of a mash. This is a gradual change and the laxative effects of the latter, place the cow in wonderful condition to calve. This I give in amounts varying with the size of the cow, but usually three pounds of bran with one-half pound of oil meal added, mixed to a wet mash will take care of the individual. At the time of partuation, usually a week afterward, I take this away from her, and, during the partuation period, which may last from six to eight hours, she receives nothing but all the warm water she can drink. At this time she should be in a well-bedded box stall that has been diinfected thoroughly before. After the calf is born the mother should be well disinfected, and the portions of her udder cleaned off well. After the calf sucks the first time, it is well to even the udder by milking a little from each quarter and relieve the udder of its strain. If the cow successfully comes through the process of partuation I place her back on the bran mash. After this, to the dry ration, and from it to the milking ration. This is a gradual process and can be governed in each case to suit the individual and surrounding conditions. The time required to get a cow on feed depends and varies upon the conditions of the cow, but usually between ten to thirty days. This is where the personal element and good judgment of the feeder comes in so successful and determines to a great extent the results that are gained in the feeding. I then increase the milking ration one pound per day until I have reached her maximum feeding capacity.

The various methods of feeding the cow depend on many different conditions. Not only does it depend upon the likes and dislikes of the individual, but upon the climate and the different kinds of crops that are grown in the different sections of the country. These conditions vary, but at the same time there is one thing that the man feeding the cow should do, and that is to arrange for the proper kind of feed in that locality, and the best quality of that feed. As the cow advances into the lactation period it is best not to let her get too low in flesh, and this is just what the good dairy cow is going to do. If the cow is milking heavy she will take a portion off her back, and to prevent too much of this, more carbonaceous feeds should be used. In case the cow shrinks in quantity and starts placing on flesh it is best to add more protein to the ration. This may be done by adding corn or hominy as in the first instance, and oil meal, cottonseed meal, or corn glutin. I prefer a nutritive ration of 1:6.5, and this will allow you to feed about one pound of protein to every four pounds of other grain in the ration. One other good factor in the ration is variety. The ration should be composed of a reasonable number of feeds, balanced so that it affords palatibility, for this alone stimulates the appetite and aids digestion. A cow will give better returns from feeds that are palatable than from those that have not been properly prepared and mixed. The number of feeds that the ration is composed of then aids in the results that are to be obtained. Frequent changes in the ration are not practical, and it causes imperfect digestion and assimilation. It is my preference to adjust the feed supply so that the rations can be made from two kinds of roughage, and an indefinite number of concentrates. If the appetizing, well-balanced ration can be fed at all times, the results obtained will be much better and the cow will be in far better physical condition. There is one method that is very good in heavy feeding of test cows, and that is to add a bran mash off and on. This does no immediate harm, does not cause the cow to lose in milk from that feeding, but does tend to break the monotony of the heavy milking ration.

Below are some of the rations that have been tried and found practical:

These may be fed according to the requirements of the individual, with all the beet pulp, clover, alfalfa and silage that she will take.

Ration No. 1—
  200 pounds of ground clipped oats.
  200 pounds of bran.
  400 pounds of yellow corn meal or corn and cob meal.
  100 pounds of cotton seed meal.
  100 pounds of oil meal, old process.

In case the oil cake is able to be used I would advise its use because of its freshness, and the palatable result that it has in the mixture.

Ration No. 2—
  200 pounds of clipped and ground oats.
  200 pounds of soft winter wheat bran.
  100 pounds of old process oil meal.
  400 pounds of ground yellow corn meal.
  200 pounds of corn glutin.
  50 pounds of salt.

Ration No. 3—
  400 pounds of corn (ground).
  300 pounds of ground oats.
  200 pounds of bran.
  200 pounds of glutin.
  200 pounds of cotton seed meal.
  200 pounds of oil meal.
  100 pounds of shorts.
  200 pounds of ground alfalfa.
  50 pounds of salt.

These rations should be made up of the best quality of grains and with enough variety to make them palatable. In making your own mixed feeds you have the opportunity of doing individual feeding and adjusting the ration to the needs of every cow in the herd. This is a factor to consider in the preservation of the udders, the life of the good cows and economy in feeding. Feeding goes hand in hand with breeding, and the success of the calves from year to year depend upon what is fed the mother.

During the lactation period the cow should be bred again, and, in case she is on test, should be bred to enter the classes set down in the rules of the American Jersey Cattle club, and it is here that the vital question arrives in the breeding game. For these classes the breeder has to make various changes in his breeding operations, and one of the dreadful results of this change is often appearing in the herd. Barren cows is often the result of this practice, and caused by neglect or having to wait to breed to enter a class. At no time do I think that a cow should go over three heat periods, and, if possible, should be bred on the third period. If this is followed, there will be less chances for nonbreeders, and I can see no reason why a regular breeding cow is not more profitable than one which has been placed in a class and becomes barren. These are assets that make up the profitable end of the herd and with the practical ideas in mind the breeder can see his way to success.

HOW THE JERSEYS LEAD

World's Champion, first five consecutive years, a Jersey—4415 lbs. butter-fat.

World's Champion, over 14 years old, a Jersey—714 lbs. butter-fat.

World's Champion for life-time production, a Jersey—7038 lbs. butter-fat.

World's Champion for reproduction and production at advanced age, a Jersey—now 22½ years old, mother of 20 heifers and 1 bull; in 24 months, starting at 18 years, she produced 800 lbs. of butter-fat and 3 living calves.

No other breed of dairy cattle has ever approached these records.

COME TO OHIO FOR GOOD JERSEYS——THREE THOUSAND BREEDERS WAITING TO SERVE YOU.

www.ingramcontent.com/pod-product-compliance
Lightning Source LLC
Chambersburg PA
CBHW082339220526
45470CB00008B/2566